Advances in Plant Biochemistry

Advances in Plant Biochemistry

Dr. R.A.S. Tomar

Editor

KOROS PRESS LIMITED

London, UK

Advances in Plant Biochemistry

© 2012
Printed in 2017 for Sale in the Indian Subcontinent

Published by
Koros Press Limited
3 The Pines, Rubery B45 9FF, Rednal,
Birmingham, United Kingdom

Tel.: +44-7826-930152
Email: info@korospress.com
www.korospress.com

ISBN: 978-1-78163-023-5

Editor: Dr. R.A.S. Tomar

Printed in UK

10 9 8 7 6 5 4 3 2 1

British Library Cataloguing in Publication Data
A CIP record for this book is available from the British Library

Exclusively distributed by CBS Publishers & Distributors Pvt. Ltd.
Sales & Distribution Rights only for India, Pakistan, Bangladesh, Sri Lanka, Nepal and Bhutan.This book is not to be sold outside these territories.

Contents

Preface

Living organisms are complex systems. Hundreds of thousands of proteins exist inside each one of us to help carry out our daily functions. These proteins are produced locally, assembled piece-by-piece to exact specifications. An enormous amount of information is required to manage this complex system correctly. This information, detailing the specific structure of the proteins inside of our bodies, is stored in a set of molecules called nucleic acids.

The most common adenosine derivative is the cyclic form, 3'-5'-cyclic adenosine monophosphate, cAMP. This compound is a very powerful second messenger involved in passing signal transduction events from the cell surface to internal proteins, e.g. cAMP-dependent protein kinase, PKA. PKA phosphorylates a number of proteins, thereby, affecting their activity either positively or negatively. Cyclic-AMP is also involved in the regulation of ion channels by direct interaction with the channel proteins, e.g. in the activation of odorant receptors by odorant molecules. Formation of cAMP occurs in response to activation of receptor coupled adenylate cyclase. These receptors can be of any type, e.g. hormone receptors or odorant receptors.

S-adenosylmethionine is a form of activated methionine which serves as a methyl donor in methylation reactions and as a source of propylamine in the synthesis of polyamines. As cells divide it is a necessity that the DNA be copied (replicated), in such a way that each daughter cell acquires the same amount of genetic material. In order for this process to proceed the two strands of the helix must first be separated, in a process termed denaturation. This process can also be carried out in vitro. If a solution of DNA is subjected to high temperature, the H-bonds between bases become unstable and the strands of the helix separate in a process of thermal denaturation.

The base composition of DNA varies widely from molecule to molecule and even within different regions of the same molecule.

Regions of the duplex that have predominantly A-T base-pairs will be less thermally stable than those rich in G-C base-pairs. In the process of thermal denaturation, a point is reached at which 50% of the DNA molecule exists as single strands. This point is the melting temperature (T_m), and is characteristic of the base composition of that DNA molecule. The T_m depends upon several factors in addition to the base composition. These include the chemical nature of the solvent and the identities and concentrations of ions in the solution. When thermally melted DNA is cooled, the complementary strands will again re-form the correct base pairs, in a process is termed annealing or hybridization. The rate of annealing is dependent upon the nucleotide sequence of the two strands of DNA.

The book is a meticulously organized and useful both for teaching and for reference. It is intended to serve plant biology and related disciplines, ranging from molecular biology and biotechnology to biochemistry.

—Editor

Biochemistry of Genome

In modern molecular biology and genetics, the genome is the entirety of an organism's hereditary information. It is encoded either in DNA or, for many types of virus, in RNA. The genome includes both the genes and the non-coding sequences of the DNA.

Origin of Term

The term was adapted in 1920 by Hans Winkler, Professor of Botany at the University of Hamburg, Germany. In Greek, the word *genome* means I become, I am born, to come into being. The Oxford English Dictionary suggests the name to be a blend of the words *gene* and *chromosome*. A few related *-ome* words already existed, such as *biome* and *rhizome*, forming a vocabulary into which *genome* fits systematically.

Overview

Some organisms have multiple copies of chromosomes, diploid, triploid, tetraploid and so on. In classical genetics, in a sexually reproducing organism (typically eukarya) the gamete has half of the number of chromosome of the somatic cell and the genome is a full set of chromosomes in a gamete. In haploid organisms, including cells of bacteria, archaea, and in organelles including mitochondria and chloroplasts, or viruses, that similarly contain genes, the single or set of circular and/or linear chains of DNA (or RNA for some viruses), likewise constitute the *genome*. The term genome can be applied specifically to mean that stored on a complete set of *nuclear DNA* (i.e., the "nuclear genome") but can also be applied to that stored within organelles that contain their own DNA, as with the "mitochondrial genome" or the "chloroplast genome". Additionally, the genome can comprise nonchromosomal genetic elements such as viruses, plasmids, and transposable elements.

When people say that the genome of a sexually reproducing species has been "sequenced", typically they are referring to a determination of the sequences of one set of autosomes and one of each type of sex chromosome, which together represent both of the possible sexes. Even in species that exist in only one sex, what is described as "a genome sequence" may be a composite read from the chromosomes of various individuals. In general use, the phrase "genetic makeup" is sometimes used conversationally to mean the genome of a particular individual or organism. The study of the global properties of genomes of related organisms is usually referred to as genomics, which distinguishes it from genetics which generally studies the properties of single genes or groups of genes.

Both the number of base pairs and the number of genes vary widely from one species to another, and there is only a rough correlation between the two (an observation known as the C-value paradox). At present, the highest known number of genes is around 60,000, for the protozoan causing trichomoniasis, almost three times as many as in the human genome.

An analogy to the human genome stored on DNA is that of instructions stored in a library:

- The library would contain 46 books (chromosomes)
- The books range in size from 400 to 3340 pages (genes)
- which is 48 to 250 million letters (A,C,G,T) per book.
- Hence the library contains over six billion letters total;
- The library fits into a cell nucleus the size of a pinpoint;
- A copy of the library (all 46 books) is contained in almost every cell of our body.

Types

Most biological entities that are more complex than a virus sometimes or always carry additional genetic material besides that which resides in their chromosomes. In some contexts, such as sequencing the genome of a pathogenic microbe, "genome" is meant to include information stored on this auxiliary material, which is carried in plasmids. In such circumstances then, "genome" describes all of the genes and information on non-coding DNA that have the potential to be present.

In eukaryotes such as plants, protozoa and animals, however, "genome" carries the typical connotation of only information on chromosomal DNA. So although these organisms contain chloroplasts and/or mitochondria that have their own DNA, the genetic information contained by DNA within these organelles is not considered part of the genome. In fact, mitochondria are sometimes said to have their own genome often referred to as the "mitochondrial genome". The DNA found within the chloroplast may be referred to as the "plastome".

Genomes and Genetic Variation

Note that a genome does not capture the genetic diversity or the genetic polymorphism of a species. For example, the human genome sequence in principle could be determined from just half the information on the DNA of one cell from one individual. To learn what variations in genetic information underlie particular traits or diseases requires comparisons across individuals. This point explains the common usage of "genome" (which parallels a common usage of "gene") to refer not to the information in any particular DNA sequence, but to a whole family of sequences that share a biological context.

Although this concept may seem counter intuitive, it is the same concept that says there is no particular shape that is the shape of a cheetah. Cheetahs vary, and so do the sequences of their genomes. Yet both the individual animals and their sequences share commonalities, so one can learn something about cheetahs and "cheetah-ness" from a single example of either.

Sequencing and Mapping

The Human Genome Project was organized to map and to sequence the human genome. Other genome projects include mouse, rice, the plant *Arabidopsis thaliana*, the puffer fish, bacteria like E. coli, etc. In 1976, Walter Fiers at the University of Ghent (Belgium) was the first to establish the complete nucleotide sequence of a viral RNA-genome (bacteriophage MS2). The first DNA-genome project to be completed was the Phage Φ-X174, with only 5386 base pairs, which was sequenced by Fred Sanger in 1977. The first bacterial genome to be completed was that of Haemophilus influenzae, completed by a team at The Institute for Genomic Research in 1995.

The development of new technologies has dramatically decreased the difficulty and cost of sequencing, and the number of complete genome sequences is rising rapidly. Among many genome database sites, the one maintained by the US National Institutes of Health is inclusive. These new technologies open up the prospect of personal genome sequencing as an important diagnostic tool. A major step toward that goal was the May 2007 *New York Times* announcement that the full genome of DNA pioneer James D. Watson was deciphered.

Whereas a genome sequence lists the order of every DNA base in a genome, a genome map identifies the landmarks. A genome map is less detailed than a genome sequence and aids in navigating around the genome.

Genome Evolution

Genomes are more than the sum of an organism's genes and have traits that may be measured and studied without reference to the details of any particular genes and their products. Researchers compare traits such as *chromosome number* (karyotype), genome size, gene order, codon usage bias, and GC-content to determine what mechanisms could have produced the great variety of genomes that exist today.

Duplications play a major role in shaping the genome. Duplications may range from extension of short tandem repeats, to duplication of a cluster of genes, and all the way to duplications of entire chromosomes or even entire genomes. Such duplications are probably fundamental to the creation of genetic novelty.

Horizontal gene transfer is invoked to explain how there is often extreme similarity between small portions of the genomes of two organisms that are otherwise very distantly related. Horizontal gene transfer seems to be common among many microbes. Also, eukaryotic cells seem to have experienced a transfer of some genetic material from their chloroplast and mitochondrial genomes to their nuclear chromosomes.

Basics of Biological Chemistry

Biochemistry, sometimes abbreviated as "BioChem", is the study of chemical processes in living organisms. Biochemistry governs all living organisms and living processes. By controlling information flow through biochemical signalling and the flow of chemical energy through metabolism; biochemical processes give rise to the seemingly magical phenomenon of life. Much of biochemistry deals with the structures and functions of cellular components such as proteins, carbohydrates, lipids, nucleic acids and other biomolecules although increasingly processes rather than individual molecules are the main focus. Over the last 40 years biochemistry has become so successful at explaining living processes that now almost all areas of the life sciences from botany to medicine are engaged in biochemical research. Today the main focus of pure biochemistry is in understanding how biological molecules give rise to the processes that occur within living cells which in turn relates greatly to the study and understanding of whole organisms.

Among the vast number of different biomolecules, many are complex and large molecules (called *polymers*), which are composed of similar repeating subunits (called *monomers*). Each class of polymeric biomolecule has a different set of subunit types. For example, a protein is a polymer whose subunits are selected from a set of 20 or more amino acids. Biochemistry studies the chemical properties of important biological molecules, like proteins, and in particular the chemistry of enzyme-catalysed reactions. The biochemistry of cell metabolism and the endocrine system has been extensively described. Other areas of biochemistry include the genetic code (DNA, RNA), protein synthesis, cell membrane transport, and signal transduction.

History

Originally, it was generally believed that life was not subject to the laws of science the way non-life was. It was thought that only living beings could produce the molecules of life (from other, previously existing biomolecules). Then, in 1828, Friedrich Wöhler published a paper on the synthesis of urea, proving that organic compounds can be created artificially.

The dawn of biochemistry may have been the discovery of the first enzyme, diastase (today called amylase), in 1833 by Anselme Payen. Eduard Buchner contributed the first demonstration of a complex biochemical process outside of a cell in 1896: alcoholic fermentation in cell extracts of yeast. Although the term "biochemistry" seems to have been first used in 1882, it is generally accepted that the formal coinage of biochemistry occurred in 1903 by Carl Neuberg, a German chemist. Previously, this area would have been referred to as physiological chemistry. Since then, biochemistry has advanced, especially since the mid-20th century, with the development of new techniques such as chromatography, X-ray diffraction, dual polarisation interferometry, NMR spectroscopy, radioisotopic labelling, electron microscopy and molecular dynamics simulations. These techniques allowed for the discovery and detailed analysis of many molecules and metabolic pathways of the cell, such as glycolysis and the Krebs cycle (citric acid cycle).

Another significant historic event in biochemistry is the discovery of the gene and its role in the transfer of information in the cell. This part of biochemistry is often called molecular biology. In the 1950s, James D. Watson, Francis Crick, Rosalind Franklin, and Maurice Wilkins were instrumental in solving DNA structure and suggesting its relationship with genetic transfer of information. In 1958, George Beadle and Edward Tatum received the Nobel Prize for work in fungi showing that one gene produces one enzyme. In 1988, Colin Pitchfork was the first person convicted of murder with DNA evidence, which led to growth of forensic science. More recently, Andrew Z. Fire and Craig C. Mello received the 2006 Nobel Prize for discovering the role of RNA interference (RNAi), in the silencing of gene expression

Today, there are three main types of biochemistry. Plant biochemistry involves the study of the biochemistry of autotrophic organisms such as photosynthesis and other plant specific biochemical processes. General biochemistry encompasses both plant and animal biochemistry. Human/medical/medicinal biochemistry focuses on the biochemistry of humans and medical illnesses.

Monomers and Polymers

The four main classes of molecules in biochemistry are carbohydrates, lipids, proteins, and nucleic acids. Many biological molecules are polymers: in this terminology, *monomers* are relatively small micromolecules that are linked together to create large macromolecules, which are known as *polymers*. When monomers are linked together to synthesize a biological polymer, they undergo a process called dehydration synthesis.

Carbohydrates

Carbohydrates are made from monomers called *monosaccharides*. Some of these monosaccharides include glucose ($C_6H_{12}O_6$), fructose ($C_6H_{12}O_6$), and deoxyribose ($C_5H_{10}O_4$). When two monosaccharides undergo dehydration synthesis, water is produced, as two hydrogen atoms and one oxygen atom are lost from the two monosaccharides' hydroxyl group.

Lipids

Lipids are usually made from one molecule of glycerol combined with other molecules. In triglycerides, the main group of bulk lipids, there is one molecule of glycerol and three fatty acids. Fatty acids are considered the monomer in that case, and may be saturated (no double bonds in the carbon chain) or unsaturated (one or more double bonds in the carbon chain). Lipids, especially phospholipids, are also used in various pharmaceutical products, either as co-solubilisers (e.g. in parenteral infusions) or else as drug carrier components (e.g. in a liposome or transfersome).

Proteins

Proteins are very large molecules – macro-biopolymers – made from monomers called *amino acids*. There are 20 standard amino acids, each containing a carboxyl group, an amino group, and a side chain (known as an "R" group). The "R" group is what makes each amino acid different, and the properties of the side chains greatly influence the overall three-dimensional conformation of a protein. When amino acids combine, they form a special bond called a peptide bond through dehydration synthesis, and become a *polypeptide*, or protein.

Nucleic Acids

Nucleic acids are the molecules that make up DNA, an extremely important substance which all cellular organisms use to store their genetic information. The most common nucleic acids are deoxyribonucleic acid and ribonucleic acid. Their monomers are called nucleotides. The most common nucleotides are adenine, cytosine, guanine, thymine, and uracil. Adenine binds with thymine and uracil; thymine only binds with adenine; and cytosine and guanine can only bind with each other.

Carbohydrates

The function of carbohydrates includes energy storage and providing structure. Sugars are carbohydrates, but not all carbohydrates are sugars. There are more carbohydrates on Earth than any other known type of biomolecule; they are used to store energy and genetic information, as well as play important roles in cell to cell interactions and communications.

Monosaccharides

Glucose

The simplest type of carbohydrate is a monosaccharide, which among other properties contains carbon, hydrogen, and oxygen, mostly in a ratio of 1:2:1 (generalized formula $C_nH_{2n}O_n$, where n is at least 3). Glucose, one of the most important carbohydrates, is an example of a monosaccharide. So is fructose, the sugar commonly associated with the sweet taste of fruits.

Some carbohydrates (especially after condensation to oligo- and polysaccharides) contain less carbon relative to H and O, which still are present in 2:1 (H:O) ratio. Monosaccharides can be grouped into aldoses (having an aldehyde group at the end of the chain, e.g. glucose) and ketoses (having a keto group in their chain; e.g. fructose). Both aldoses and ketoses occur in an equilibrium (starting with chain lengths of C4) cyclic forms. These are generated by bond formation between one of the hydroxyl groups of the sugar chain with the carbon of the aldehyde or keto group to form a hemiacetal bond. This leads to saturated five-membered (in furanoses) or six-membered (in pyranoses) heterocyclic rings containing one O as heteroatom.

Disaccharides

Figure: ordinary sugar and probably the most familiar carbohydrate.

Two monosaccharides can be joined together using dehydration synthesis, in which a hydrogen atom is removed from the end of one molecule and a hydroxyl group (—OH) is removed from the other; the remaining residues are then attached at the sites from which the atoms were removed. The H—OH or H_2O is then released as a molecule of water, hence the term *dehydration*. The new molecule, consisting of two monosaccharides, is called a *disaccharide* and is conjoined together by a glycosidic or ether bond.

The reverse reaction can also occur, using a molecule of water to split up a disaccharide and break the glycosidic bond; this is termed *hydrolysis*. The most well-known disaccharide is sucrose, ordinary sugar (in scientific contexts, called *table sugar* or *cane sugar* to differentiate it from other sugars). Sucrose consists of a glucose molecule and a fructose molecule joined together. Another important disaccharide is lactose, consisting of a glucose molecule and a galactose molecule. As most humans age, the production of lactase, the enzyme that hydrolyzes lactose back into glucose and galactose, typically decreases. This results in lactase deficiency, also called *lactose intolerance.*

Sugar polymers are characterised by having reducing or non-reducing ends. A reducing end of a carbohydrate is a carbon atom which can be in equilibrium with the open-chain aldehyde or keto form.

If the joining of monomers takes place at such a carbon atom, the free hydroxy group of the pyranose or furanose form is exchanged with an OH-side chain of another sugar, yielding a full acetal.

This prevents opening of the chain to the aldehyde or keto form and renders the modified residue non-reducing. Lactose contains a reducing end at its glucose moiety, whereas the galactose moiety form a full acetal with the C4-OH group of glucose. Saccharose does not have a reducing end because of full acetal formation between the aldehyde carbon of glucose (C1) and the keto carbon of fructose (C2).

Oligosaccharides and Polysaccharides

Figure: Cellulose as polymer of β-D-glucose

When a few (around three to six) monosaccharides are joined together, it is called an *oligosaccharide*. These molecules tend to be used as markers and signals, as well as having some other uses. Many monosaccharides joined together make a polysaccharide. They can be joined together in one long linear chain, or they may be branched. Two of the most common polysaccharides are cellulose and glycogen, both consisting of repeating glucose monomers.

- *Cellulose* is made by plants and is an important structural component of their cell walls. Humans can neither manufacture nor digest it.

- *Glycogen*, on the other hand, is an animal carbohydrate; humans and other animals use it as a form of energy storage.

Use of Carbohydrates as an Energy Source

Glucose is the major energy source in most life forms. For instance, polysaccharides are broken down into their monomers (glycogen phosphorylase removes glucose residues from glycogen). Disaccharides like lactose or sucrose are cleaved into their two component monosaccharides.

Glycolysis (Anaerobic)

Glucose is mainly metabolized by a very important ten-step pathway called glycolysis, the net result of which is to break down one molecule of glucose into two molecules of pyruvate; this also produces a net two molecules of ATP, the energy currency of cells, along with two reducing equivalents in the form of converting NAD^+ to NADH. This does not require oxygen; if no oxygen is available (or the cell cannot use oxygen), the NAD is restored by converting the pyruvate to lactate (lactic acid) (e.g. in humans) or to ethanol plus carbon dioxide (e.g. in yeast). Other monosaccharides like galactose and fructose can be converted into intermediates of the glycolytic pathway.

Aerobic

In aerobic cells with sufficient oxygen, like most human cells, the pyruvate is further metabolized. It is irreversibly converted to acetyl-CoA, giving off one carbon atom as the waste product carbon dioxide, generating another reducing equivalent as NADH. The two molecules acetyl-CoA (from one molecule of glucose) then enter the citric acid cycle, producing two more molecules of ATP, six more NADH molecules and two reduced (ubi)quinones (via $FADH_2$ as enzyme-bound cofactor), and releasing the remaining carbon atoms as carbon dioxide.

The produced NADH and quinol molecules then feed into the enzyme complexes of the respiratory chain, an electron transport system transferring the electrons ultimately to oxygen and conserving the released energy in the form of a proton gradient over a membrane (inner mitochondrial membrane in eukaryotes). Thereby, oxygen is reduced to water and the original electron acceptors NAD^+ and quinone are regenerated. This is why humans breathe in oxygen and breathe out carbon dioxide. The energy released from transferring the electrons from high-energy states in NADH and quinol is conserved first as proton gradient and converted to ATP via ATP synthase. This generates an additional *28* molecules of ATP (24 from the 8 NADH + 4 from the 2 quinols), totalling to 32 molecules of ATP conserved per degraded glucose (two from glycolysis + two from the citrate cycle). It is clear that using oxygen to completely oxidize glucose provides an organism with far more energy than any oxygen-independent metabolic feature, and this is thought to be the reason why complex life appeared only after Earth's atmosphere accumulated large amounts of oxygen.

Gluconeogenesis

In vertebrates, vigorously contracting skeletal muscles (during weightlifting or sprinting, for example) do not receive enough oxygen to meet the energy demand, and so they shift to anaerobic metabolism, converting glucose to lactate. The liver regenerates the glucose, using a process called gluconeogenesis. This process is not quite the opposite of glycolysis, and actually requires three times the amount of energy gained from glycolysis (six molecules of ATP are used, compared to the two gained in glycolysis). Analogous to the above reactions, the glucose produced can then undergo glycolysis in tissues that need energy, be stored as glycogen (or starch in plants), or be converted to other monosaccharides or joined into di- or oligosaccharides. The combined pathways of glycolysis during exercise, lactate's crossing via the bloodstream to the liver, subsequent gluconeogenesis and release of glucose into the bloodstream is called the Cori cycle.

Proteins

Like carbohydrates, some proteins perform largely structural roles. For instance, movements of the proteins actin and myosin ultimately are responsible for the contraction of skeletal muscle. One property many proteins have is that they specifically bind to a certain molecule or class of molecules—they may be *extremely* selective in what they bind. Antibodies are an example of proteins that attach to one specific type of molecule.

In fact, the enzyme-linked immunosorbent assay (ELISA), which uses antibodies, is currently one of the most sensitive tests modern medicine uses to detect various biomolecules. Probably the most important proteins, however, are the enzymes. These molecules recognize specific reactant molecules called *substrates*; they then catalyse the reaction between them. By lowering the activation energy, the enzyme speeds up that reaction by a rate of [11]or more: a reaction that would normally take over 3,000 years to complete spontaneously might take less than a second with an enzyme. The enzyme itself is not used up in the process, and is free to catalyse the same reaction with a new set of substrates. Using various modifiers, the activity of the enzyme can be regulated, enabling control of the biochemistry of the cell as a whole.

In essence, proteins are chains of amino acids. An amino acid consists of a carbon atom bound to four groups. One is an amino group, $-NH_2$, and one is a carboxylic acid group, $-COOH$ (although these exist as $-NH_3^+$ and $-COO^-$ under physiologic conditions). The third is a simple hydrogen atom. The fourth is commonly denoted "$-R$" and is different for each amino acid. There are twenty standard amino acids. Some of these have functions by themselves or in a modified form; for instance, glutamate functions as an important neurotransmitter.

Amino acids can be joined together via a peptide bond. In this dehydration synthesis, a water molecule is removed and the peptide bond connects the nitrogen of one amino acid's amino group to the carbon of the other's carboxylic acid group. The resulting molecule is called a *dipeptide*, and short stretches of amino acids (usually, fewer than around thirty) are called *peptides* or polypeptides. Longer stretches merit the title *proteins*. As an example, the important blood serum protein albumin contains 585 amino acid residues.

The structure of proteins is traditionally described in a hierarchy of four levels. The primary structure of a protein simply consists of its linear sequence of amino acids; for instance, "alanine-glycine-tryptophan-serine-glutamate-asparagine-glycine-lysine-...". Secondary structure is concerned with local morphology (morphology being the study of structure). Some combinations of amino acids will tend to curl up in a coil called an α-helix or into a sheet called a β-sheet; some α-helixes can be seen in the hemoglobin schematic above. Tertiary structure is the entire three-dimensional shape of the protein. This shape is determined by the sequence of amino acids. In fact, a single change can change the entire structure. The alpha chain of hemoglobin contains 146 amino acid residues; substitution of the glutamate residue at position 6 with a valine residue changes the behaviour of hemoglobin so much that it results in sickle-cell disease. Finally quaternary structure is concerned with the structure of a protein with multiple peptide subunits, like hemoglobin with its four subunits. Not all proteins have more than one subunit.

Ingested proteins are usually broken up into single amino acids or dipeptides in the small intestine, and then absorbed. They can then be joined together to make new proteins. Intermediate products of glycolysis, the citric acid cycle, and the pentose phosphate pathway can be used to make all twenty amino acids, and most bacteria and plants possess all the necessary enzymes to synthesize them.

Humans and other mammals, however, can only synthesize half of them. They cannot synthesize isoleucine, leucine, lysine, methionine, phenylalanine, threonine, tryptophan, and valine. These are the essential amino acids, since it is essential to ingest them. Mammals do possess the enzymes to synthesize alanine, asparagine, aspartate, cysteine, glutamate, glutamine, glycine, proline, serine, and tyrosine, the nonessential amino acids. While they can synthesize arginine and histidine, they cannot produce it in sufficient amounts for young, growing animals, and so these are often considered essential amino acids.

If the amino group is removed from an amino acid, it leaves behind a carbon skeleton called an α-keto acid. Enzymes called transaminases can easily transfer the amino group from one amino acid (making it an α-keto acid) to another α-keto acid (making it an amino acid). This is important in the biosynthesis of amino acids, as for many of the pathways, intermediates from other biochemical pathways are converted to the α-keto acid skeleton, and then an amino group is added, often via transamination. The amino acids may then be linked together to make a protein.

A similar process is used to break down proteins. It is first hydrolysed into its component amino acids. Free ammonia (NH_3), existing as the ammonium ion (NH_4^+) in blood, is toxic to life forms. A suitable method for excreting it must therefore exist. Different strategies have evolved in different animals, depending on the animals' needs. Unicellular organisms, of course, simply release the ammonia into the environment. Similarly, bony fish can release the ammonia into the water where it is quickly diluted. In general, mammals convert the ammonia into urea, via the urea cycle.

Lipids

The term lipid comprises a diverse range of molecules and to some extent is a catchall for relatively water-insoluble or nonpolar compounds of biological origin, including waxes, fatty acids, fatty-acid derived phospholipids, sphingolipids, glycolipids and terpenoids (e.g. retinoids and steroids). Some lipids are linear aliphatic molecules, while others have ring structures. Some are aromatic, while others are not. Some are flexible, while others are rigid.

Most lipids have some polar character in addition to being largely nonpolar. Generally, the bulk of their structure is nonpolar or hydrophobic ("water-fearing"), meaning that it does not interact well with polar solvents like water. Another part of their structure is polar or hydrophilic ("water-loving") and will tend to associate with polar solvents like water.

This makes them amphiphilic molecules (having both hydrophobic and hydrophilic portions). In the case of cholesterol, the polar group is a mere -OH (hydroxyl or alcohol). In the case of phospholipids, the polar groups are considerably larger and more polar, as described below.

Lipids are an integral part of our daily diet. Most oils and milk products that we use for cooking and eating like butter, cheese, ghee etc., are composed of fats. Vegetable oils are rich in various polyunsaturated fatty acids (PUFA). Lipid-containing foods undergo digestion within the body and are broken into fatty acids and glycerol, which are the final degradation products of fats and lipids.

Nucleic Acids

A nucleic acid is a complex, high-molecular-weight biochemical macromolecule composed of nucleotide chains that convey genetic information. The most common nucleic acids are deoxyribonucleic acid (DNA) and ribonucleic acid (RNA). Nucleic acids are found in all living cells and viruses. Aside from the genetic material of the cell, nucleic acids often play a role as second messengers, as well as forming the base molecule for adenosine triphosphate, the primary energy-carrier molecule found in all living organisms.

Nucleic acid, so called because of its prevalence in cellular nuclei, is the generic name of the family of biopolymers. The monomers are called nucleotides, and each consists of three components: a nitrogenous heterocyclic base (either a purine or a pyrimidine), a pentose sugar, and a phosphate group.

Different nucleic acid types differ in the specific sugar found in their chain (e.g. DNA or deoxyribonucleic acid contains 2-deoxyriboses). Also, the nitrogenous bases possible in the two nucleic acids are different: adenine, cytosine, and guanine occur in both RNA and DNA, while thymine occurs only in DNA and uracil occurs in RNA.

Relationship to other "Molecular-scale" Biological Sciences

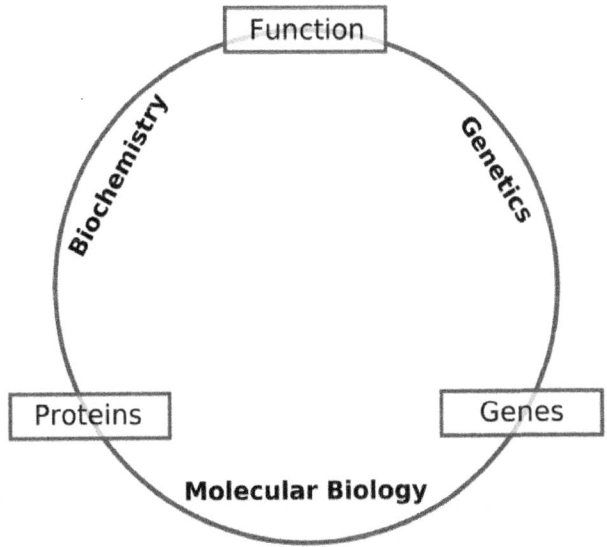

Figure: Schematic relationship between biochemistry, genetics and molecular biology

Researchers in biochemistry use specific techniques native to biochemistry, but increasingly combine these with techniques and ideas from genetics, molecular biology and biophysics. There has never been a hard-line between these disciplines in terms of content and technique. Today the terms *molecular biology* and *biochemistry* are nearly interchangeable. The schematic that depicts one possible view of the relationship between the fields:

- *Biochemistry* is the study of the chemical substances and vital processes occurring in living organisms. Biochemists focus heavily on the role, function, and structure of biomolecules. The study of the chemistry behind biological processes and the synthesis of biologically active molecules are examples of biochemistry.

- *Genetics* is the study of the effect of genetic differences on organisms. Often this can be inferred by the absence of a normal component (e.g. one gene). The study of "mutants" –

organisms which lack one or more functional components with respect to the so-called "wild type" or normal phenotype. Genetic interactions (epistasis) can often confound simple interpretations of such "knock-out" studies.

· *Molecular biology* is the study of molecular underpinnings of the process of replication, transcription and translation of the genetic material. The central dogma of molecular biology where genetic material is transcribed into RNA and then translated into protein, despite being an oversimplified picture of molecular biology, still provides a good starting point for understanding the field. This picture, however, is undergoing revision in light of emerging novel roles for RNA.

· *Chemical Biology* seeks to develop new tools based on small molecules that allow minimal perturbation of biological systems while providing detailed information about their function. Further, chemical biology employs biological systems to create non-natural hybrids between biomolecules and synthetic devices (for example emptied viral capsids that can deliver gene therapy or drug molecules).

2

Recent Developments in Biochemistry

Genetic Engineering

Genetic engineering, also called genetic modification, is the direct human manipulation of an organism's genetic material in a way that does not occur under natural conditions. It involves the use of recombinant DNA techniques, but does not include traditional animal and plant breeding or mutagenesis. Any organism that is generated using these techniques is considered to be a genetically modified organism. The first organisms genetically engineered were bacteria in 1973 and then mice in 1974. Insulin producing bacteria were commercialized in 1982 and genetically modified food has been sold since 1994.

The most common form of genetic engineering involves the insertion of new genetic material at an unspecified location in the host genome. This is accomplished by isolating and copying the genetic material of interest, generating a construct containing all the genetic elements for correct expression, and then inserting this construct into the host organism. Other forms of genetic engineering include gene targeting and knocking out specific genes via engineered nucleases such as zinc finger nucleases or engineered homing endonucleases.

Genetic engineering techniques have been applied in numerous fields including research, biotechnology, and medicine. Medicines such as insulin and human growth hormone are now produced in bacteria, experimental mice such as the oncomouse and the knockout mouse are being used for research purposes and insect resistant and/or herbicide tolerant crops have been commercialized. Genetically engineered plants and animals capable of producing biotechnology drugs more cheaply than current methods (called pharming) are also being developed and in 2009 the FDA approved the sale of the pharmaceutical protein antithrombin produced in the milk of genetically engineered goats.

Definition

Genetic engineering alters the genetic makeup of an organism using techniques that introduce heritable material prepared outside the organism either directly into the host or into a cell that is then fused or hybridized with the host. This involves using recombinant nucleic acid (DNA or RNA) techniques to form new combinations of heritable genetic material followed by the incorporation of that material either indirectly through a vector system or directly through micro-injection, macro-injection and micro-encapsulation techniques.

Genetic engineering does not include traditional animal and plant breeding, in vitro fertilisation, induction of polyploidy, mutagenesis and cell fusion techniques that do not use recombinant nucleic acids or a genetically modified organism in the process. Cloning and stem cell research, although not considered genetic engineering, are closely related and genetic engineering can be used within them. Synthetic biology is an emerging discipline that takes genetic engineering a step further by introducing artificially synthesized genetic material from raw materials into an organism.

If genetic material from another species is added to the host, the resulting organism is called transgenic. If genetic material from the same species or a species that can naturally breed with the host is used the resulting organism is called cisgenic. Genetic engineering can also be used to remove genetic material from the target organism, creating a knock out organism. In Europe genetic modification is synonymous with genetic engineering while within the United States of America it can also refer to conventional breeding methods.

History

Humans have altered the genomes of species for thousands of years through artificial selection and more recently mutagenesis. Genetic engineering as the direct manipulation of DNA by humans outside breeding and mutations has only existed since the 1970s. The term "genetic engineering" was first coined by Jack Williamson in his science fiction novel *Dragon's Island*, published in 1951, one year before DNA's role in heredity was confirmed by Alfred Hershey and Martha Chase, and two years before James Watson and Francis Crick showed that the DNA molecule has a double-helix structure.

In 1972 Paul Berg created the first recombinant DNA molecules by combined DNA from the monkey virus SV40 with that of the lambda virus. In 1973 Herbert Boyer and Stanley Cohen created the first transgenic organism by inserting antibiotic resistance genes into the plasmid of an *E. coli* bacterium. A year later Rudolf Jaenisch created a transgenic mouse by introducing foreign DNA into its embryo, making it the world's first transgenic animal.

In 1976 Genentech, the first genetic engineering company was founded by Herbert Boyer and Robert Swanson and a year later and the company produced a human protein (somatostatin) in *E.coli*. Genentech announced the production of genetically engineered human insulin in 1978. In 1980, the U.S. Supreme Court in the Diamond v. Chakrabarty case ruled that genetically altered life could be patented. The insulin produced by bacteria, branded humulin, was approved for release by the Food and Drug Administration in 1982.

The first field trials of genetically engineered plants occurred in France and the USA in 1986, tobacco plants were engineered to be resistant to herbicides. The People's Republic of China was the first country to commercialize transgenic plants, introducing a virus-resistant tobacco in 1992. In 1994 Calgene attained approval to commercially release the Flavr Savr tomato, a tomato engineered to have a longer shelf life.

In 1994, the European Union approved tobacco engineered to be resistant to the herbicide bromoxynil, making it the first genetically engineered crop commercialized in Europe. In 1995, Bt Potato was approved safe by the Environmental Protection Agency, making it the first pesticide producing crop to be approved in the USA. In 2009 11 transgenic crops were grown commercially in 25 countries, the largest of which by area grown were the USA, Brazil, Argentina, India, Canada, China, Paraguay and South Africa.

In 2010, scientists at the J. Craig Venter Institute, announced that they had created the first synthetic bacterial genome, and added it to a cell containing no DNA. The resulting bacterium, named Synthia, was the world's first synthetic life form.

Process

Isolating the Gene

First, the gene to be inserted into the genetically modified organism must be chosen and isolated. Presently, most genes transferred into plants provide protection against insects or tolerance to herbicides. In animals the majority of genes used are growth hormone genes. Once chosen the genes must be isolated. This typically involves multiplying the gene using polymerase chain reaction (PCR). If the chosen gene or the donor organism's genome has been well studied it may be present in a genetic library. If the DNA sequence is known, but no copies of the gene are available, it can be artificially synthesized. Once isolated, the gene is inserted into a bacterial plasmid.

Constructs

The gene to be inserted into the genetically modified organism must be combined with other genetic elements in order for it to work properly. The gene can also be modified at this stage for better expression or effectiveness. As well as the gene to be inserted most constructs contain a promoter and terminator region as well as a selectable marker gene. The promoter region initiates transcription of the gene and can be used to control the location and level of gene expression, while the terminator region ends transcription. The selectable marker, which in most cases confers antibiotic resistance to the organism it is expressed in, is needed to determine which cells are transformed with the new gene. The constructs are made using recombinant DNA techniques, such as restriction digests, ligations and molecular cloning.

Gene Targeting

The most common form of genetic engineering involves inserting new genetic material randomly within the host genome. Other techniques allow new genetic material to be inserted at a specific location in the host genome or generate mutations at desired genomic loci capable of knocking out endogenous genes. The technique of gene targeting uses homologous recombination to target desired changes to a specific endogenous gene. This tends to occur at a relatively low frequency in plants and animals and generally requires the use of selectable markers. The frequency of gene targeting can be greatly enhanced with the use of engineered nucleases such as zinc finger nucleases, engineered homing endonucleases, or nucleases created from TAL effectors. In addition to enhancing gene targeting, engineered nucleases can also be used to introduce mutations at endogenous genes that generate a gene knockout.

Transformation

About 1% of bacteria are naturally able to take up foreign DNA but it can also be induced in other bacteria. Stressing the bacteria for example, with a heat shock or an electric shock, can make the cell membrane permeable to DNA that may then incorporate into their genome or exist as extrachromosomal DNA. DNA is generally inserted into animal cells using microinjection, where it can be injected through the cells nuclear envelope directly into the nucleus or through the use of viral vectors. In plants the DNA is generally inserted using *Agrobacterium*-mediated recombination or biolistics.

In *Agrobacterium*-mediated recombination the plasmid construct must also contain T-DNA. *Agrobacterium* naturally inserts DNA from a tumour inducing plasmid into any susceptible plant's genome it infects, causing crown gall disease. The T-DNA region of this plasmid is responsible for insertion of the DNA. The genes to be inserted are cloned into a binary vector, which contains T-DNA and can be grown in both *E. Coli* and *Agrobacterium*. Once the binary vector is constructed the plasmid is transformed into *Agrobacterium* containing no plasmids and plant cells are infected. The *Agrobacterium* will then naturally insert the genetic material into the plant cells.

In biolistics particles of gold or tungsten are coated with DNA and then shot into young plant cells or plant embryos. Some genetic material will enter the cells and transform them. This method can be used on plants that are not susceptible to *Agrobacterium* infection and also allows transformation of plant plastids. Another transformation method for plant and animal cells is electroporation. Electroporation involves subjecting the plant or animal cell to an electric shock, which can make the cell membrane permeable to plasmid DNA. In some cases the electroporated cells will incorporate the DNA into their genome. Due to the damage caused to the cells and DNA the transformation efficiency of biolistics and electroporation is lower than agrobacterial mediated transformation and microinjection.

Selection

Not all the organism's cells will be transformed with the new genetic material; in most cases a selectable marker is used to differentiate transformed from untransformed cells. If a cell has been successfully transformed with the DNA it will also contain the marker gene. By growing the cells in the presence of an antibiotic or chemical that selects or marks the cells expressing that gene it is possible to separate the transgenic events from the non-transgenic. Another method of screening involves using a DNA probe that will only stick to the inserted gene. A number of strategies have been developed that can remove the selectable marker from the mature transgenic plant.

Regeneration

As often only a single cell is transformed with genetic material the organism must be regrown from that single cell. As bacteria consist of a single cell and reproduce clonally regeneration is not necessary. In plants this is accomplished through the use of tissue culture. Each plant species has different requirements for successful regeneration through tissue culture. If successful an adult plant is produced that contains the transgene in every cell. In animals it is necessary to ensure that the inserted DNA is present in the embryonic stem cells. When the offspring is produced they can be screened for the presence of the gene. All offspring from the first generation will be heterozygous for the inserted gene and must be mated together to produce a homozygous animal.

Confirmation

Further tests using PCR, Southern Blots and Bioassays are needed to confirm that the gene is expressed and functions correctly. The organism's offspring are also tested to ensure that the trait can be inherited and that it follows a Mendelian inheritance pattern.

Applications

Genetic engineering has applications in medicine, research, industry and agriculture and can be used on a wide range of plants, animals and micro organism.

Medicine

In medicine genetic engineering has been used to mass produce insulin, human growth hormones, follistim (for treating infertility), human albumin, monoclonal antibodies, antihemophilic factors, vaccines and many other drugs. Vaccination generally involves injecting weak live, killed or inactivated forms of viruses or their toxins into the person being immunized. Genetically engineered viruses are being developed that can still confer immunity, but lack the infectious sequences. Mouse hybridomas, cells fused together to create monoclonal antibodies, have been humanised through genetic engineering to create human monoclonal antibodies.

Genetic engineering is used to create animal models of human diseases. Genetically modified mice are the most common genetically engineered animal model. They have been used to study and model cancer (the oncomouse), obesity, heart disease, diabetes, arthritis, substance abuse, anxiety, aging and Parkinson disease. Potential cures can be tested against these mouse models. Also genetically modified pigs have been bred with the aim of increasing the success of pig to human organ transplantation.

Gene therapy is the genetic engineering of humans by replacing defective human genes with functional copies. This can occur in somatic tissue or germline tissue. If the gene is inserted into the germline tissue it can be passed down to that person's descendants. Gene therapy has been used to treat patients suffering from immune deficiencies (notably Severe combined immunodeficiency) and trials have been carried out on other genetic disorders. The success of gene therapy so far has been limited and a patient (Jesse Gelsinger) has died during a clinical trial testing a new treatment. There are also ethical concerns should the technology be used not just for treatment, but for enhancement, modification or alteration of a human beings' appearance, adaptability, intelligence, character or behaviour. The distinction between cure and enhancement can also be difficult to establish. Transhumanists consider the enhancement of humans desirable.

Research

Genetic engineering is an important tool for natural scientists. Genes and other genetic information from a wide range of organisms are transformed into bacteria for storage and modification, creating genetically modified bacteria in the process. Bacteria are cheap, easy to grow, clonal, multiply quickly, relatively easy to transform and can be stored at -80°C almost indefinitely. Once a gene is isolated it can be stored inside the bacteria providing an unlimited supply for research.

Organisms are genetically engineered to discover the functions of certain genes. This could be the effect on the phenotype of the organism, where the gene is expressed or what other genes it interacts with. These experiments generally involve loss of function, gain of function, tracking and expression.

- Loss of function experiments, such as in a gene knockout experiment, in which an organism is engineered to lack the activity of one or more genes. A knockout experiment involves the creation and manipulation of a DNA construct *in vitro*, which, in a simple knockout, consists of a copy of the desired gene, which has been altered such that it is non-functional. Embryonic stem cells incorporate the altered gene, which replaces the already present functional copy. These stem cells are injected into blastocysts, which are implanted into surrogate mothers. This allows the experimenter to analyze the defects caused by this mutation and thereby determine the role of particular genes. It is used especially frequently in developmental biology. Another method, useful in organisms such as Drosophila (fruit fly), is to induce mutations in a large population and then screen the progeny for the desired mutation. A similar process can be used in both plants and prokaryotes.

- Gain of function experiments, the logical counterpart of knockouts. These are sometimes performed in conjunction with knockout experiments to more finely establish the function of the desired gene. The process is much the same as that in knockout engineering, except that the construct is designed to increase the function of the gene, usually by providing extra copies of the gene or inducing synthesis of the protein more frequently.

- Tracking experiments, which seek to gain information about the localization and interaction of the desired protein. One way to do this is to replace the wild-type gene with a 'fusion' gene, which is a juxtaposition of the wild-type gene with a reporting element such as green fluorescent protein (GFP) that will allow easy visualization of the products of the genetic modification. While this is a useful technique, the manipulation can destroy the function of the gene, creating secondary effects and possibly calling into question the results of the experiment. More sophisticated techniques are now in development that can track protein products without mitigating their function, such as the addition of small sequences that will serve as binding motifs to monoclonal antibodies.

- Expression studies aim to discover where and when specific proteins are produced. In these experiments, the DNA sequence before the DNA that codes for a protein, known as a gene's promoter, is reintroduced into an organism with the protein coding region replaced by a reporter gene such as GFP or an enzyme that catalyzes the production of a dye. Thus the time and place where a particular protein is produced can be observed. Expression studies can be taken a step further by altering the promoter to find which pieces are crucial for the proper expression of the gene and are actually bound by transcription factor proteins; this process is known as promoter bashing.

Industrial

By engineering genes into bacterial plasmids it is possible to create a biological factory that can produce proteins and enzymes. Some genes do not work well in bacteria, so yeast, a eukaryote, can also be used. Bacteria and yeast factories have been used to produce medicines such as insulin, human growth hormone, and vaccines, supplements such as tryptophan, aid in the production of food (chymosin in cheese making) and fuels. Other applications involving genetically engineered bacteria being investigated involve making the bacteria perform tasks outside their natural cycle, such as cleaning up oil spills, carbon and other toxic waste.

Agriculture

One of the best-known and controversial applications of genetic engineering is the creation of genetically modified food. There are three generations of genetically modified crops. First generation crops have been commercialized and most provide protection from insects and/or resistance to herbicides. There are also fungal and virus resistant crops developed or in development. They have been developed to make the insect and weed management of crops easier and can indirectly increase crop yield. The second generation of genetically modified crops being developed aim to directly improve yield by improving salt, cold or drought tolerance and to increase the nutritional value of the crops. The third generation consists of pharmaceutical crops, crops that contain edible vaccines and other drugs. Some agriculturally important animals have been genetically modified with growth hormones to increase their size while others have been engineered to express drugs and other proteins in their milk.

The genetic engineering of agricultural crops can increase the growth rates and resistance to different diseases caused by pathogens and parasites. This is beneficial as it can greatly increase the production of food sources with the usage of fewer resources that would be required to host the world's growing populations. These modified crops would also reduce the usage of chemicals, such as fertilizers and pesticides, and therefore decrease the severity and frequency of the damages produced by these chemical pollution.

Ethical and safety concerns have been raised around the use of genetically modified food. A major safety concern relates to the human health implications of eating genetically modified food, in particular whether toxic or allergic reactions could occur. Gene flow into related non-transgenic crops, off target effects on beneficial organisms and the impact on biodiversity are important environmental issues. Ethical concerns involve religious issues, corporate control of the food supply, intellectual property rights and the level of labelling needed on genetically modified products.

Other Uses

In materials science, a genetically modified virus has been used to construct a more environmentally friendly lithium-ion battery. Some bacteria have been genetically engineered to create black and white photographs while others have potential to be used as sensors by expressing a fluorescent protein under certain environmental conditions. Genetic engineering is also being used to create BioArt and novelty items such as blue roses, and glowing fish.

Opposition and Criticism

A 2010 study of Canola found transgenes in 80% of wild (uncultivated or "feral") varieties in North Dakota, meaning 80% of the plants which had established themselves in the area were genetically engineered varieties. The researchers stated that "we found the highest densities of [such transgene-containing] plants near agricultural fields and along major freeways, but we were also finding plants in the middle of nowhere" adding that "over time,..the build-up of different types of herbicide resistance in feral [natural] canola and closely related weeds, like field mustard, could make it more difficult to manage these plants using herbicides."

Pharmacogenomics

What is Pharmacogenomics?

Pharmacogenomics is the study of how an individual's genetic inheritance affects the body's response to drugs. The term comes from the words pharmacology and genomics and is thus the intersection of pharmaceuticals and genetics. Pharmacogenomics holds the promise that drugs might one day be tailor-made for individuals and adapted to each person's own genetic makeup. Environment, diet, age, lifestyle, and state of health all can influence a person's response to medicines, but understanding an individual's genetic makeup is thought to be the key to creating personalized drugs with greater efficacy and safety. Pharmacogenomics combines traditional pharmaceutical sciences such as biochemistry with annotated knowledge of genes, proteins, and single nucleotide polymorphisms.

What are the Anticipated Benefits of Pharmacogenomics?

More Powerful Medicines: Pharmaceutical companies will be able to create drugs based on the proteins, enzymes, and RNA molecules associated with genes and diseases. This will facilitate drug discovery and allow drug makers to produce a therapy more targeted to specific diseases. This accuracy not only will maximize therapeutic effects but also decrease damage to nearby healthy cells.

Better, Safer Drugs the First Time: Instead of the standard trial-and-error method of matching patients with the right drugs, doctors will be able to analyze a patient's genetic profile and prescribe the best available drug therapy from the beginning. Not only will this take the guesswork out of finding the right drug, it will speed recovery time and increase safety as the likelihood of adverse reactions is eliminated. Pharmacogenomics has the potential to dramatically reduce the the estimated 100,000 deaths and 2 million hospitalizations that occur each year in the United States as the result of adverse drug response.

More Accurate Methods of Determining Appropriate Drug Dosages: Current methods of basing dosages on weight and age will be replaced with dosages based on a person's genetics —how well the body processes the medicine and the time it takes to metabolize it. This will maximize the therapy's value and decrease the likelihood of overdose.

Advanced Screening for Disease: Knowing one's genetic code will allow a person to make adequate lifestyle and environmental changes at an early age so as to avoid or lessen the severity of a genetic disease. Likewise, advance knowledge of a particular disease susceptibility will allow careful monitoring, and treatments can be introduced at the most appropriate stage to maximize their therapy.

Better Vaccines: Vaccines made of genetic material, either DNA or RNA, promise all the benefits of existing vaccines without all the risks. They will activate the immune system but will be unable to cause infections. They will be inexpensive, stable, easy to store, and capable of being engineered to carry several strains of a pathogen at once.

Improvements in the Drug Discovery and Approval Process: Pharmaceutical companies will be able to discover potential therapies more easily using genome targets. Previously failed drug candidates may be revived as they are matched with the niche population they serve. The drug approval process should be facilitated as trials are targeted for specific genetic population groups —providing greater degrees of success. The cost and risk of clinical trials will be reduced by targeting only those persons capable of responding to a drug.

Decrease in the Overall Cost of Health Care: Decreases in the number of adverse drug reactions, the number of failed drug trials, the time it takes to get a drug approved, the length of time patients are on medication, the number of medications patients must take to find an effective therapy, the effects of a disease on the body (through early detection), and an increase in the range of possible drug targets will promote a net decrease in the cost of health care.

Is Pharmacogenomics in Use Today?

To a limited degree. The cytochrome P450 (CYP) family of liver enzymes is responsible for breaking down more than 30 different classes of drugs. DNA variations in genes that code for these enzymes can influence their ability to metabolize certain drugs. Less active or inactive forms of CYP enzymes that are unable to break down and efficiently eliminate drugs from the body can cause drug overdose in patients. Today, clinical trials researchers use genetic tests for variations in cytochrome P450 genes to screen and monitor patients. In addition, many pharmaceutical companies screen their chemical compounds to see how well they are broken down by variant forms of CYP enzymes.

Another enzyme called TPMT (thiopurine methyltransferase) plays an important role in the chemotherapy treatment of a common childhood leukemia by breaking down a class of therapeutic compounds called thiopurines. A small percentage of Caucasians have genetic variants that prevent them from producing an active form of this protein. As a result, thiopurines elevate to toxic levels in the patient because the inactive form of TMPT is unable to break down the drug. Today, doctors can use a genetic test to screen patients for this deficiency, and the TMPT activity is monitored to determine appropriate thiopurine dosage levels.

What are some of the Barriers to Pharmacogenomics Progress?

Pharmacogenomics is a developing research field that is still in its infancy. Several of the following barriers will have to be overcome before many pharmacogenomics benefits can be realized.

- Complexity of finding gene variations that affect drug response -Single nucleotide polymorphisms (SNPs) are DNA sequence variations that occur when a single nucleotide (A,T,C,or G) in the genome sequence is altered. SNPs occur every 100 to 300 bases along the 3-billion-base human genome, therefore millions of SNPs must be identified and analysed to determine their involvement (if any) in drug response. Further complicating the process is our limited knowledge of which genes are involved with each drug response. Since many genes are likely to influence responses, obtaining the big picture on the impact of gene variations is highly time-consuming and complicated.

- Limited drug alternatives -Only one or two approved drugs may be available for treatment of a particular condition. If patients have gene variations that prevent them using these drugs, they may be left without any alternatives for treatment.

- Disincentives for drug companies to make multiple pharmacogenomic products -Most pharmaceutical companies have been successful with their "one size fits all" approach to drug development. Since it costs hundreds of millions of dollars to bring a drug to market, will these companies be willing to develop alternative drugs that serve only a small portion of the population?

- Educating healthcare providers -Introducing multiple pharmacogenomic products to treat the same condition for different population subsets undoubtedly will complicate the process of prescribing and dispensing drugs. Physicians must execute an extra diagnostic step to determine which drug is best suited to each patient. To interpret the diagnostic accurately and recommend the best course of treatment for each patient, all prescribing physicians, regardless of specialty, will need a better understanding of genetics.

Molecules to Medicines

As you've read so far, the most important goals of modern pharmacology are also the most obvious. Pharmacologists want to design, and be able to produce in sufficient quantity, drugs that will act in a specific way without too many side effects. They also want to deliver the correct amount of a drug to the proper place in the body. But turning molecules into medicines is more easily said than done. Scientists struggle to fulfill the twin challenges of drug design and drug delivery.

Medicine Hunting

While sometimes the discovery of potential medicines falls to researchers' good luck, most often pharmacologists, chemists, and other scientists looking for new drugs plod along methodically for years, taking suggestions from nature or clues from knowledge about how the body works.

National Agriculture Library, ARS, USDA

Finding chemicals' cellular targets can educate scientists about how drugs work. Aspirin's molecular target, the enzyme cyclooxygenase, or COX, was discovered this way in the early 1970s in Nobel Prize-winning work by pharmacologist John Vane, then at the Royal College of Surgeons in London, England. Another example is colchicine, a relatively old drug that is still widely used to treat gout, an excruciatingly painful type of arthritis in which needle-like crystals of uric acid clog joints, leading to swelling, heat, pain, and stiffness.

Lab experiments with colchicine led scientists to this drug's molecular target, a cell-scaffolding protein called tubulin. Colchicine works by attaching itself to tubulin, causing certain parts of a cell's architecture to crumble, and this action can interfere with a cell's ability to move around. Researchers suspect that in the case of gout, colchicine works by halting the migration of immune cells called granulocytes that are responsible for the inflammation characteristic of gout.

Current estimates indicate that scientists have identified roughly 500 to 600 molecular targets where medicines may have effects in the body. Medicine hunters can strategically "discover" drugs by designing molecules to "hit" these targets. That has already happened in some cases. Researchers knew just what they were looking for when they designed the successful AIDS drugs called HIV protease inhibitors.

Previous knowledge of the three-dimensional structure of certain HIV proteins (the target) guided researchers to develop drugs shaped to block their action. Protease inhibitors have extended the lives of many people with AIDS.

However, sometimes even the most targeted approaches can end up in big surprises. The New York City pharmaceutical firm Pfizer had a blood pressure-lowering drug in mind, when instead its scientists discovered Viagra®, a best-selling drug approved to treat erectile dysfunction. Initially, researchers had planned to create a heart drug, using knowledge they had about molecules that make blood clot and molecular signals that instruct blood vessels to relax. What the scientists did not know was how their candidate drug would fare in clinical trials.

Sildenafil (Viagra's chemical name) did not work very well as a heart medicine, but many men who participated in the clinical testing phase of the drug noted one side effect in particular: erections. Viagra works by boosting levels of a natural molecule called cyclic GMP that plays a key role in cell signalling in many body tissues. This molecule does a good job of opening blood vessels in the penis, leading to an erection.

A Drug by Another Name

As pet owners know, you *can* teach some old dogs new tricks. In a similar vein, scientists have in some cases found new uses for "old" drugs. Remarkably, the potential new uses often have little in common with a drug's product label (its "old" use). For example, chemist Eric Oldfield of the University of Illinois at Urbana-Champaign discovered that one class of drugs called bisphosphonates, which are currently approved to treat osteoporosis and other bone disorders, may also be useful for treating malaria, Chagas' disease, leishmaniasis, and AIDS-related infections like toxoplasmosis.

Previous research by Oldfield and his coworkers had hinted that the active ingredient in the bisphosphonate medicines Fosamax®, Actonel®, and Aredia® blocks a critical step in the metabolism of parasites, the microorganisms that cause these diseases. To test whether this was true, Oldfield gave the medicines to five different types of parasites, each grown along with human cells in a plastic lab dish. The scientists found that small amounts of the osteoporosis drugs killed the parasites while sparing human cells. The researchers are now testing the drugs in animal models of the parasitic diseases and so far have obtained cures—in mice—of certain types of leishmaniasis. If these studies prove that bisphosphonate drugs work in larger animal models, the next step will be to find out if the medicines can thwart these parasitic diseases in humans.

21st-Century Science

While strategies such as chemical genetics can quicken the pace of drug discovery, other approaches may help expand the number of molecular targets from several hundred to several thousand. Many of these new avenues of research hinge on biology.

Relatively new brands of research that are stepping onto centre stage in 21st-century science include genomics (the study of all of an organism's genetic material), proteomics (the study of all of an organism's proteins), and bioinformatics (using computers to sift through large amounts of biological data). The "omics" revolution in biomedicine stems from biology's gradual transition from a gathering, descriptive enterprise to a science that will someday be able to model and predict biology. If you think 25,000 genes is a lot (the number of genes in the human genome), realize that each gene can give rise to different variations of the same protein, each with a different molecular job. Scientists estimate that humans have hundreds of thousands of protein variants. Clearly, there's lots of work to be done, which will undoubtedly keep researchers busy for years to come.

A Chink in Cancer's Armor

Recently, researchers made an exciting step forward in the treatment of cancer. Years of basic research investigating circuits of cellular communication led scientists to tailor-make a new kind of cancer medicine. In May 2001, the drug Gleevec™ was approved to treat a rare cancer of the blood called chronic myelogenous leukemia (CML). The Food and Drug Administration described Gleevec's approval as "...a testament to the groundbreaking scientific research taking place in labs throughout America."

Researchers designed this drug to halt a cell-communication pathway that is always "on" in CML. Their success was founded on years of experiments in the basic biology of how cancer cells grow. The discovery of Gleevec is an example of the success of so-called molecular targeting: understanding how diseases arise at the level of cells, then figuring out ways to treat them. Scores of drugs, some to treat cancer but also many other health conditions, are in the research pipeline as a result of scientists' eavesdropping on how cells communicate.

Rush Delivery

Finding new medicines and cost-effective ways to manufacture them is only half the battle. An enormous challenge for pharmacologists is figuring out how to get drugs to the right place, a task known as drug delivery. Ideally, a drug should enter the body, go directly to the diseased site while bypassing healthy tissue, do its job, and then disappear. Unfortunately, this rarely happens with the typical methods of delivering drugs: swallowing and injection. When swallowed, many medicines made of protein are never absorbed into the bloodstream because they are quickly chewed up by enzymes as they pass through the digestive system. If the drug does get to the blood from the intestines, it falls prey to liver enzymes. For doctors prescribing such drugs, this first-pass effect means that several doses of an oral drug are needed before enough makes it to the blood. Drug injections also cause problems, because they are expensive, difficult for patients to self-administer, and are unwieldy if the drug must be taken daily. Both methods of administration also result in fluctuating levels of the drug in the blood, which is inefficient and can be dangerous.

What to do? Pharmacologists can work around the first-pass effect by delivering medicines via the skin, nose, and lungs. Each of these methods bypasses the intestinal tract and can increase the amount of drug getting to the desired site of action in the body. Slow, steady drug delivery directly to the bloodstream—without stopping at the liver first—is the primary benefit of skin patches, which makes this form of drug delivery particularly useful when a chemical must be administered over a long period.

Hormones such as testosterone, progesterone, and estrogen are available as skin patches. These forms of medicines enter the blood via a meshwork of small arteries, veins, and capillaries in the skin. Researchers also have developed skin patches for a wide variety of other drugs. Some of these include Duragesic® (a prescription-only pain medicine), Transderm Scop® (a motion-sickness drug), and Transderm Nitro® (a blood vessel-widening drug used to treat chest pain associated with heart disease). Despite their advantages, however, skin patches have a significant drawback. Only very small drug molecules can get into the body through the skin.

Inhaling drugs through the nose or mouth is another way to rapidly deliver drugs and bypass the liver. Inhalers have been a mainstay of asthma therapy for years, and doctors prescribe nasal steroid drugs for allergy and sinus problems. Researchers are investigating insulin powders that can be inhaled by people with diabetes who rely on insulin to control their blood sugar daily. This still-experimental technology stems from novel uses of chemistry and engineering to manufacture insulin particles of just the right size. Too large, and the insulin particles could lodge in the lungs; too small, and the particles will be exhaled. If clinical trials with inhaled insulin prove that it is safe and effective, then this therapy could make life much easier for people with diabetes.

Reading a Cell MAP

Scientists try hard to listen to the noisy, garbled "discussions" that take place inside and between cells. Less than a decade ago, scientists identified one very important cellular communication stream called MAP (mitogen-activated protein) kinase signalling. Today, molecular pharmacologists such as Melanie H. Cobb of the University of Texas Southwestern Medical Centre at Dallas are studying how MAP kinase signalling pathways malfunction in unhealthy cells.

Some of the interactions between proteins in these pathways involve adding and taking away tiny molecular labels called phosphate groups. Kinases are the enzymes that add phosphate groups to proteins, and this process is called phosphorylation. Marking proteins in this way assigns the proteins a code, instructing the cell to do something, such as divide or grow. The body employs many, many signalling pathways involving hundreds of different kinase enzymes. Some of the important functions performed by MAP kinase pathways include instructing immature cells how to "grow up" to be specialized cell types like muscle cells, helping cells in the pancreas respond to the hormone insulin, and even telling cells how to die.

Since MAP kinase pathways are key to so many important cell processes, researchers consider them good targets for drugs. Clinical trials are under way to test various molecules that, in animal studies, can effectively lock up MAP kinase signalling when it's not wanted, for example, in cancer and in diseases involving an overactive immune system, such as arthritis. Researchers predict that if drugs to block MAP kinase signalling prove effective in people, they will likely be used in combination with other medicines that treat a variety of health conditions, since many diseases are probably caused by simultaneous errors in multiple signalling pathways.

Transportation Dilemmas

Scientists are solving the dilemma of drug delivery with a variety of other clever techniques. Many of the techniques are geared toward sneaking through the cellular gate-keeping systems' membranes. The challenge is a chemistry problem—most drugs are water-soluble, but membranes are oily. Water and oil don't mix, and thus many drugs can't enter the cell. To make matters worse, size matters too. Membranes are usually constructed to permit the entry of only small nutrients and hormones, often through private cellular alleyways called transporters.

Many pharmacologists are working hard to devise ways to work not against, but *with* nature, by learning how to hijack molecular transporters to shuttle drugs into cells. Gordon Amidon, a pharmaceutical chemist at the University of Michigan-Ann Arbor, has been studying one particular transporter in mucosal membranes lining the digestive tract. The transporter, called hPEPT1, normally serves the body by ferrying small, electrically charged particles and small protein pieces called peptides into and out of the intestines.

Amidon and other researchers discovered that certain medicines, such as the antibiotic penicillin and certain types of drugs used to treat high blood pressure and heart failure, also travel into the intestines via hPEPT1. Recent experiments revealed that the herpes drug Valtrex® and the AIDS drug Retrovir® also hitch a ride into intestinal cells using the hPEPT1 transporter. Amidon wants to extend this list by synthesizing hundreds of different molecules and testing them for their ability to use hPEPT1 and other similar transporters. Recent advances in molecular biology, genomics, and bioinformatics have sped the search for molecules that Amidon and other researchers can test.

Scientists are also trying to slip molecules through membranes by cloaking them in disguise. Steven Regen of Lehigh University in Bethlehem, Pennsylvania, has manufactured miniature chemical umbrellas that close around and shield a molecule when it encounters a fatty membrane and then spread open in the watery environment inside a cell. So far, Regen has only used test molecules, not actual drugs, but he has succeeded in getting molecules that resemble small segments of DNA across membranes. The ability to do this in humans could be a crucial step in successfully delivering therapeutic molecules to cells via gene therapy.

Act Like a Membrane

Researchers know that high concentrations of chemotherapy drugs will kill every single cancer cell growing in a lab dish, but getting enough of these powerful drugs to a tumour in the body without killing too many healthy cells along the way has been exceedingly difficult. These powerful drugs can do more harm than good by severely sickening a patient during treatment.

Lawrence Mayer, Ludger Ickenstein, Katrina Edwards

Some researchers are using membrane-like particles called liposomes to package and deliver drugs to tumours. Liposomes are oily, microscopic capsules that can be filled with biological cargo, such as a drug. They are very, very small—only one one-thousandth the width of a single human hair. Researchers have known about liposomes for many years, but getting them to the right place in the body hasn't been easy. Once in the bloodstream, these foreign particles are immediately shipped to the liver and spleen, where they are destroyed.

Materials engineer David Needham of Duke University in Durham, North Carolina, is investigating the physics and chemistry of liposomes to better understand how the liposomes and their cancer-fighting cargo can travel through the body. Needham worked for 10 years to create a special kind of liposome that melts at just a few degrees above body temperature. The end result is a tiny molecular "soccer ball" made from two different oils that wrap around a drug. At room temperature, the liposomes are solid and they stay solid at body temperature, so they can be injected into the bloodstream. The liposomes are designed to spill their drug cargo into a tumour when heat is applied to the cancerous tissue. Heat is known to perturb tumours, making the blood vessels surrounding cancer cells extra-leaky. As the liposomes approach the warmed tumour tissue, the "stitches" of the miniature soccer balls begin to dissolve, rapidly leaking the liposome's contents.

Needham and Duke oncologist Mark Dewhirst teamed up to do animal studies with the heat-activated liposomes. Experiments in mice and dogs revealed that, when heated, the drug-laden capsules flooded tumours with a chemotherapy drug and killed the cancer cells inside. Researchers hope to soon begin the first stage of human studies testing the heat-triggered liposome treatment in patients with prostate and breast cancer. The results of these and later clinical trials will determine whether liposome therapy can be a useful weapon for treating breast and prostate cancer and other hard-to-treat solid tumours.

3

Chemistry of Life

Basic Chemistry of Nucleic Acid

Living organisms are complex systems. Hundreds of thousands of proteins exist inside each one of us to help carry out our daily functions. These proteins are produced locally, assembled piece-by-piece to exact specifications. An enormous amount of information is required to manage this complex system correctly.

This information, detailing the specific structure of the proteins inside of our bodies, is stored in a set of molecules called nucleic acids.

Adenosine Derivatives

The most common adenosine derivative is the cyclic form, 3'-5'-cyclic adenosine monophosphate, cAMP. This compound is a very powerful second messenger involved in passing signal transduction events from the cell surface to internal proteins, e.g. cAMP-dependent protein kinase, PKA. PKA phosphorylates a number of proteins, thereby, affecting their activity either positively or negatively.

Cyclic-AMP is also involved in the regulation of ion channels by direct interaction with the channel proteins, e.g. in the activation of odorant receptors by odorant molecules. Formation of cAMP occurs in response to activation of receptor coupled adenylate cyclase. These receptors can be of any type, e.g. hormone receptors or odorant receptors.

S-adenosylmethionine is a form of activated methionine which serves as a methyl donor in methylation reactions and as a source of propylamine in the synthesis of polyamines.

methionine + ATP $\xrightarrow{\text{methionine adenosyltransferase}}$ Pp$_i$ + P$_i$

Structure and Synthesis of S-adenosylmethionine.

Guanosine Derivatives

A cyclic form of GMP (cGMP) also is found in cells involved as a second messenger molecule. In many cases its' role is to antagonize the effects of cAMP. Formation of cGMP occurs in response to receptor mediated signals similar to those for activation of adenylate cyclase. However, in this case it is guanylate cyclase that is coupled to the receptor.

The most important cGMP coupled signal transduction cascade is that photoreception. However, in this case activation of rhodopsin (in the rods) or other opsins (in the cones) by the absorption of a photon of light (through 11-*cis*-retinal covalently associated with rhodopsin and opsins) activates transducin which in turn activates a cGMP specific phosphodiesterase that hydrolyzes cGMP to GMP. This lowers the effective concentration of cGMP bound to gated ion channels resulting in their closure and a concomitant hyperpolarization of the cell.

Nucleotide Derivatives in tRNAs

As many as 25% of the nucleotides found in tRNAs are modified post-transcriptionally. At least 80 different modified nucleotides have been identified at >60 different positions in various tRNAs.

The two most commonly occurring modified nucleotides are dihydrouridine (abbreviated D) and pseudouridine (abbreviated Ψ) each of which is found in a characteristic loop structure in tRNAs. Dihydrouridine is found in the D-loop (hence the name of the loop) and pseudouridine is found in the TΨC loop (hence the name of the loop.

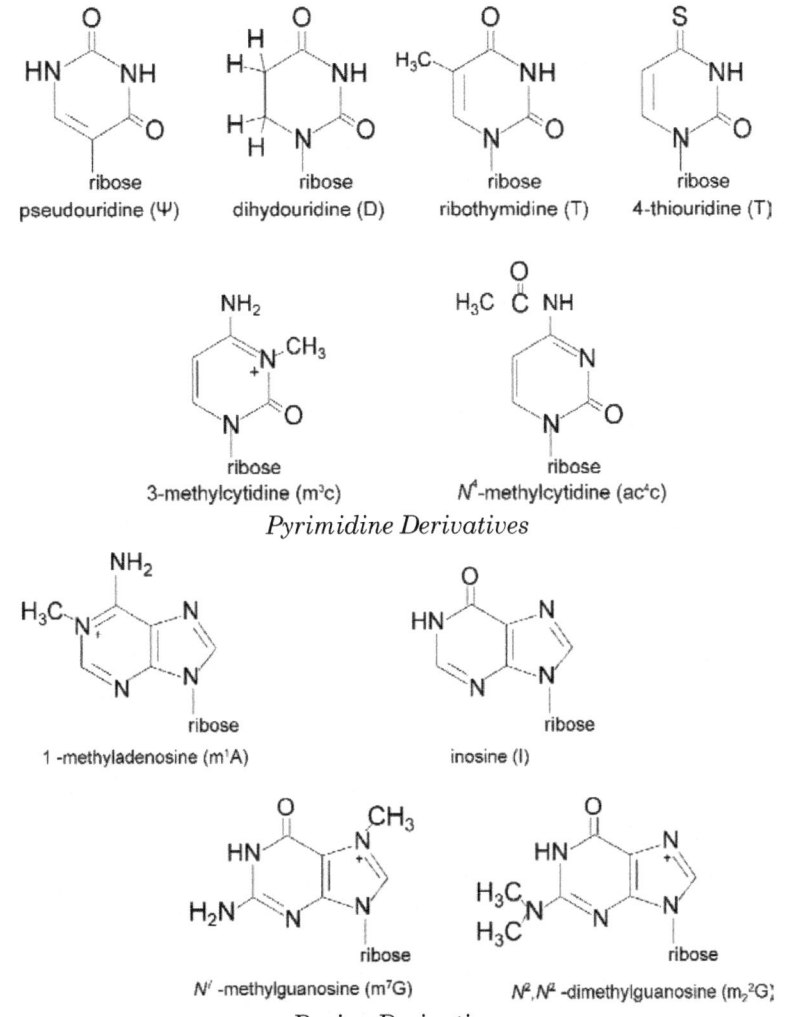

Pyrimidine Derivatives

Purine Derivatives

Synthetic Nucleotide Analogs

Many nucleotide analogues are chemically synthesized and used for their therapeutic potential. The nucleotide analogues can be utilized to inhibit specific enzymatic activities. A large family of analogues are used as anti-tumour agents, for instance, because they interfere with the synthesis of DNA and thereby preferentially kill rapidly dividing cells such as tumour cells. Some of the nucleotide analogues commonly used in chemotherapy are 6-mercaptopurine, 5-fluorouracil, 5-iodo-2'-deoxyuridine and 6-thioguanine. Each of these compounds disrupts the normal replication process by interfering with the formation of correct Watson-Crick base-pairing.

Nucleotide analogs also have been targeted for use as antiviral agents. Several analogs are used to interfere with the replication of HIV, such as AZT (azidothymidine) and ddI (dideoxyinosine).

Several purine analogs are used to treat gout. The most common is allopurinol, which resembles hypoxanthine. Allopurinol inhibits the activity of xanthine oxidase, an enzyme involved in *de novo* purine biosynthesis. Additionally, several nucleotide analogues are used after organ transplantation in order to suppress the immune system and reduce the likelihood of transplant rejection by the host.

Polynucleotides

Polynucleotides are formed by the condensation of two or more nucleotides. The condensation most commonly occurs between the alcohol of a 5'-phosphate of one nucleotide and the 3'-hydroxyl of a second, with the elimination of H_2O, forming a phosphodiester bond. The formation of phosphodiester bonds in DNA and RNA exhibits directionality. The primary structure of DNA and RNA (the linear arrangement of the nucleotides) proceeds in the 5'—>3' direction. The common representation of the primary structure of DNA or RNA molecules is to write the nucleotide sequences from left to right synonymous with the 5'—>3' direction as shown:

5'-pGpApTpC-3'

Structure of DNA

Utilizing X-ray diffraction data, obtained from crystals of DNA, James Watson and Francis Crick proposed a model for the structure of DNA. This model (subsequently verified by additional data) predicted that DNA would exist as a helix of two complementary antiparallel strands, wound around each other in a rightward direction and stabilized by H-bonding between bases in adjacent strands. In the Watson-Crick model, the bases are in the interior of the helix aligned at a nearly 90 degree angle relative to the axis of the helix. Purine bases form hydrogen bonds with pyrimidines, in the crucial phenomenon of base pairing. Experimental determination has shown that, in any given molecule of DNA, the concentration of adenine (A) is equal to thymine (T) and the concentration of cytidine (C) is equal to guanine (G). This means that A will only base-pair with T, and C with G. According to this pattern, known as Watson-Crick base-pairing, the base-pairs composed of G and C contain three H-bonds, whereas those of A and T contain two H-bonds. This makes G-C base-pairs more stable than A-T base-pairs.

A   T

A-T Base Pair

G   C

G-C Base Pair

The antiparallel nature of the helix stems from the orientation of the individual strands. From any fixed position in the helix, one strand is oriented in the 5'—>3' direction and the other in the 3'—>5' direction. On its exterior surface, the double helix of DNA contains two deep grooves between the ribose-phosphate chains. These two grooves are of unequal size and termed the major and minor grooves. The difference in their size is due to the asymmetry of the deoxyribose rings and the structurally distinct nature of the upper surface of a base-pair relative to the bottom surface.

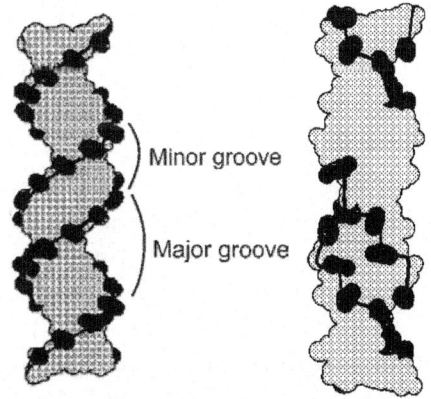

Minor groove

Major groove

Structure of B-DNA   *Structure of Z-DNA*

The double helix of DNA has been shown to exist in several different forms, depending upon sequence content and ionic conditions of crystal preparation.

The B-form of DNA prevails under physiological conditions of low ionic strength and a high degree of hydration. Regions of the helix that are rich in pCpG dinucleotides can exist in a novel left-handed helical conformation termed Z-DNA. This conformation results from a 180 degree change in the orientation of the bases relative to that of the more common A- and B-DNA.

Table: Parameters of Major DNA Helices

Parameters	A Form	B Form	Z-Form
Direction of helical rotation	Right	Right	Left
Residues per turn of helix	11	10	12 base pairs
Rotation of helix per residue (in degrees)	33	36	-30
Base tilt relative to helix axis (in degrees)	20	6	7
Major groove	narrow and deep	wide and deep	Flat
Minor groove	wide and shallow	narrow and deep	narrow and deep
Orientation of N-glycosidic bond	anti	anti	anti for Pyrimidines, syn for Purines
Comments		most prevalent within cells	occurs in stretches of alternating purine-pyrimidine base pairs

Thermal Properties of DNA

As cells divide it is a necessity that the DNA be copied (replicated), in such a way that each daughter cell acquires the same amount of genetic material. In order for this process to proceed the two strands of the helix must first be separated, in a process termed denaturation. This process can also be carried out in vitro. If a solution of DNA is subjected to high temperature, the H-bonds between bases become unstable and the strands of the helix separate in a process of thermal denaturation.

The base composition of DNA varies widely from molecule to molecule and even within different regions of the same molecule. Regions of the duplex that have predominantly A-T base-pairs will be less thermally stable than those rich in G-C base-pairs. In the process of thermal denaturation, a point is reached at which 50% of the DNA molecule exists as single strands. This point is the melting temperature (T_m), and is characteristic of the base composition of that DNA molecule. The T_m depends upon several factors in addition to the base composition. These include the chemical nature of the solvent and the identities and concentrations of ions in the solution.

When thermally melted DNA is cooled, the complementary strands will again re-form the correct base pairs, in a process is termed annealing or hybridization. The rate of annealing is dependent upon the nucleotide sequence of the two strands of DNA.

Analysis of DNA Structure

Chromatography: Several of the chromatographic techniques available for the characterization of proteins can also be applied to the characterization of DNA. The most commonly used technique is HPLC (high performance liquid chromatography). Affinity chromatographic techniques also can be employed. One common affinity matrix is hydroxyapatite (a form of calcium phosphate), which binds double-stranded DNA with a higher affinity than single-stranded DNA.

Electrophoresis: This procedure can serve the same function with regard to DNA molecules as it does for the analysis of proteins. However, since DNA molecules have much higher molecular weights than proteins, the molecular sieve used in electrophoresis of DNA must be different as well. The material of choice is agarose, a carbohydrate polymer purified from a salt water algae. It is a copolymer of mannose and galactose that when melted and re-cooled forms a gel with pores sizes dependent upon the concentration of agarose. The phosphate backbone of DNA is highly negatively charged, therefore DNA will migrate in an electric field. The size of DNA fragments can then be determined by comparing their migration in the gel to known size standards. Extremely large molecules of DNA (in the range of 30kb–10Mb) are effectively separated in agarose gels using pulsed-field gel electrophoresis (PFGE). This technique employs two or more electrodes, placed orthogonally with respect to the gel, that receive short alternating pulses of current. PFGE allows whole chromosomes and large portions of chromosomes to be analysed.

Carbohydrates

In many ways, our bodies can be thought of as chemical processing plants. Chemicals are taken in, processed through various types of reactions, and then distributed throughout the body to be used immediately or stored for later use. The chemicals used by the body can be divided into two broad categories: macronutrients, those substances that we need to eat regularly in fairly large quantities, and micronutrients, those substances that we need only in small amounts. Three major classes of macronutrients are essential to living organisms: carbohydrates, fats, and proteins

A carbohydrate is an organic compound which has the empirical formula $C_m(H_2O)_n$; that is, consists only of carbon, hydrogen and oxygen, with a hydrogen:oxygen atom ratio of 2:1 (as in water). Carbohydrates can be viewed as hydrates of carbon, hence their name. Structurally however, it is more accurate to view them as polyhydroxy aldehydes and ketones.

The term is most common in biochemistry, where it is a synonym of saccharide. The carbohydrates (saccharides) are divided into four chemical groupings: monosaccharides, disaccharides, oligosaccharides, and polysaccharides. In general, the monosaccharides and disaccharides, which are smaller (lower molecular weight) carbohydrates, are commonly referred to as sugars. While the scientific nomenclature of carbohydrates is complex, the names of the monosaccharides and disaccharides very often end in the suffix-ose. For example, blood sugar is the monosaccharide glucose, table sugar is the disaccharide sucrose, and milk sugar is the disaccharide lactose.

Carbohydrates perform numerous roles in living things. Polysaccharides serve for the storage of energy (e.g., starch and glycogen) and as structural components (e.g., cellulose in plants and chitin in arthropods). The 5-carbon monosaccharide ribose is an important component of coenzymes (e.g., ATP, FAD, and NAD) and the backbone of the genetic molecule known as RNA. The related deoxyribose is a component of DNA. Saccharides and their derivatives include many other important biomolecules that play key roles in the immune system, fertilization, preventing pathogenesis, blood clotting, and development.

In food science and in many informal contexts, the term carbohydrate often means any food that is particularly rich in the complex carbohydrate starch (such as cereals, bread and pasta) or simple carbohydrates, such as sugar (found in candy, jams and desserts).

Structure

Formerly the name "carbohydrate" was used in chemistry for any compound with the formula $C_m(H_2O)_n$. Following this definition, some chemists considered formaldehyde CH_2O to be the simplest carbohydrate, while others claimed that title for glycolaldehyde Today the term is generally understood in the biochemistry sense, which excludes compounds with only one or two carbons.

Natural saccharides are generally built of simple carbohydrates called monosaccharides with general formula $(CH_2O)_n$ where n is three or more. A typical monosaccharide has the structure H-$(CHOH)_x(C=O)-(CHOH)_y$-H, that is, an aldehyde or ketone with many hydroxyl groups added, usually one on each carbon atom that is not part of the aldehyde or ketone functional group. Examples of monosaccharides are glucose, fructose, and glyceraldehyde. However, some biological substances commonly called "monosaccharides" do not conform to this formula (e.g., uronic acids and deoxy-sugars such as fucose), and there are many chemicals that do conform to this formula but are not considered to be monosaccharides (e.g., formaldehyde CH_2O and inositol $(CH_2O)_6$). The open-chain form of a monosaccharide often coexists with a closed ring form where the aldehyde/ketone carbonyl group carbon (C=O) and hydroxyl group (-OH) react forming a hemiacetal with a new C-O-C bridge.

Monosaccharides can be linked together into what are called polysaccharides (or oligosaccharides) in a large variety of ways. Many carbohydrates contain one or more modified monosaccharide units that have had one or more groups replaced or removed. For example, deoxyribose, a component of DNA, is a modified version of ribose; chitin is composed of repeating units of N-acetylglucosamine, a nitrogen-containing form of glucose.

Monosaccharides

Monosaccharides are the most basic units of biologically important carbohydrates. They are the simplest form of sugar and are usually colorless, water-soluble, crystalline solids. Some monosaccharides have a sweet taste. Examples of monosaccharides include glucose (dextrose), fructose (levulose), galactose, xylose and ribose. Monosaccharides are the building blocks of disaccharides such as sucrose and polysaccharides (such as cellulose and starch). Further, each carbon atom that supports a hydroxyl group (except for the first and last) is chiral, giving rise to a number of isomeric forms all with the same chemical formula. For instance, galactose and glucose are both aldohexoses, but have different chemical and physical properties.

Structure

With few exceptions (e.g., deoxyribose), monosaccharides have the chemical formula $C_x(H_2O)_y$ with the chemical structure $H(CHOH)_nC=O(CHOH)_mH$. If n or m is zero, it is an aldehyde and is termed an aldose; otherwise, it is a ketone and is termed a ketose. Monosaccharides contain either a ketone or aldehyde functional group, and hydroxyl groups on most or all of the non-carbonyl carbon atoms.

Monosaccharide Momenclature

Monosaccharides are classified by the number of carbon atoms they contain:

- Triose, 3 carbon atoms
- Tetrose, 4 carbon atoms
- Pentose, 5 carbon atoms
- Hexose, 6 carbon atoms
- Heptose, 7 carbon atoms.

Some monosaccharides exist in equilibrium between its open chain and the cyclic hemiacetal forms:

Aldoses can form:

- Pyranose results from the intramolecular nucleophilic addition of the C5-OH group to the C1 carbonyl group, yielding a 6-membered ring.

Ketoses can form:

- Pyranose results from the intramolecular nucleophilic addition of the -OH at C6 to the C1 carbonyl group, forming a 6-membered ring.
- Furanose results from the intramolecular nucleophilic addition of the C4-OH group at C5 and to the C1 carbonyl group, yielding a 5-membered ring.

Monosaccharides are classified by the type of carbonyl group they contain:

- Aldose, -CHO (aldehyde)
- Ketose, C=O (ketone).

Isomerism

The total number of possible stereoisomers of one compound (n) is dependent on the number of stereogenic centres (c) in the molecule. The upper limit for the number of possible stereoisomers is $n = 2^c$. The only monosaccharide without an isomer is dihydroxyacetone or DHA.

Monosaccharides are classified according to their molecular configuration at the chiral carbon furthest removed from the aldehyde or ketone group. The chirality at this carbon is compared to the chirality of carbon 2 on glyceraldehyde. If it is equivalent to D-glyceraldehyde' s C2, the sugar is D; if it is equivalent to L-glyceraldehyde's C2, the sugar is L. Due to the chirality of the sugar molecules, an aqueous solution of a D or L saccharides will rotate light. D-glyceraldehyde causes polarized light to rotate clockwise (dextrorotary); L-glyceraldehyde causes polarized light to rotate counterclockwise (levorotary). Unlike glyceraldehyde, D/L designation on more complex sugars is not associated with their direction of light rotation. Since more complex sugars contain multiple chiral carbons, the direction of light rotation cannot be predicted by the chirality of the carbon that defines D/L nomenclature.

- D, configuration as in D-glyceraldehyde
- L, configuration as in L-glyceraldehyde.

All these classifications can be combined, resulting in names like *D-aldohexose* or *ketotriose*.

Derivatives

A large number of biologically important modified monosaccharides exist:

- Amino sugars such as:
  — Galactosamine
  — Glucosamine
  — Sialic acid
  — *N*-Acetylglucosamine
- Sulfosugars such as:
  — Sulfoquinovose.

Monosaccharides are the simplest carbohydrates in that they cannot be hydrolysed to smaller carbohydrates. They are aldehydes or ketones with two or more hydroxyl groups. The general chemical formula of an unmodified monosaccharide is $(C \cdot H_2O)_n$, literally a "carbon hydrate." Monosaccharides are important fuel molecules as well as building blocks for nucleic acids. The smallest monosaccharides, for which n = 3, are dihydroxyacetone and D- and L-glyceraldehyde.

Classification of monosaccharides

Monosaccharides are classified according to three different characteristics: the placement of its carbonyl group, the number of carbon atoms it contains, and its chiral handedness. If the carbonyl group is an aldehyde, the monosaccharide is an aldose; if the carbonyl group is a ketone, the monosaccharide is a ketose. Monosaccharides with three carbon atoms are called trioses, those with four are called tetroses, five are called pentoses, six are hexoses, and so on. These two systems of classification are often combined. For example, glucose is an aldohexose (a six-carbon aldehyde), ribose is an aldopentose (a five-carbon aldehyde), and fructose is a ketohexose (a six-carbon ketone).

Each carbon atom bearing a hydroxyl group (-OH), with the exception of the first and last carbons, are asymmetric, making them stereocenters with two possible configurations each (R or S). Because of this asymmetry, a number of isomers may exist for any given monosaccharide formula. The aldohexose D-glucose, for example, has the formula $(C \cdot H_2O)_6$, of which all but two of its six carbons atoms are stereogenic, making D-glucose one of $2^4 = 16$ possible stereoisomers. In the case of glyceraldehyde, an aldotriose, there is one pair of possible stereoisomers, which are enantiomers and epimers. 1,3-dihydroxyacetone, the ketose corresponding to the aldose glyceraldehyde, is a symmetric molecule with no stereocenters). The assignment of D or L is made according to the orientation of the asymmetric carbon furthest from the carbonyl group: in a standard Fischer projection if the hydroxyl group is on the right the molecule is a D sugar, otherwise it is an L sugar. The "D-" and "L-" prefixes should not be confused with "d-" or "l-", which indicate the direction that the sugar rotates plane polarized light. This usage of "d-" and "l-" is no longer followed in carbohydrate chemistry.

Ring-straight chain isomerism

The aldehyde or ketone group of a straight-chain monosaccharide will react reversibly with a hydroxyl group on a different carbon atom to form a hemiacetal or hemiketal, forming a heterocyclic ring with an oxygen bridge between two carbon atoms. Rings with five and six atoms are called furanose and pyranose forms, respectively, and exist in equilibrium with the straight-chain form.

During the conversion from straight-chain form to cyclic form, the carbon atom containing the carbonyl oxygen, called the anomeric carbon, becomes a stereogenic centre with two possible configurations: The oxygen atom may take a position either above or below the plane of the ring. The resulting possible pair of stereoisomers are called anomers. In the *α anomer*, the -OH substituent on the anomeric carbon rests on the opposite side (trans) of the ring from the CH_2OH side branch. The alternative form, in which the CH_2OH substituent and the anomeric hydroxyl are on the same side (cis) of the plane of the ring, is called the *β anomer*. You can remember that the β anomer is cis by the mnemonic, "It's always better to βe up". Because the ring and straight-chain forms readily interconvert, both anomers exist in equilibrium. In a Fischer Projection, the α anomer is represented with the anomeric hydroxyl group *trans* to the CH_2OH and *cis* in the β anomer.

Use in living organisms

Monosaccharides are the major source of fuel for metabolism, being used both as an energy source (glucose being the most important in nature) and in biosynthesis. When monosaccharides are not immediately needed by many cells they are often converted to more space efficient forms, often polysaccharides. In many animals, including humans, this storage form is glycogen, especially in liver and muscle cells. In plants, starch is used for the same purpose.

Disaccharides

Two joined monosaccharides are called a disaccharide and these are the simplest polysaccharides. Examples include sucrose and lactose. They are composed of two monosaccharide units bound together by a covalent bond known as a glycosidic linkage formed via a dehydration reaction, resulting in the loss of a hydrogen atom from one monosaccharide and a hydroxyl group from the other. The formula of unmodified disaccharides is $C_{12}H_{22}O_{11}$. Although there are numerous kinds of disaccharides, a handful of disaccharides are particularly notable. Sucrose, pictured to the right, is the most abundant disaccharide, and the main form in which carbohydrates are transported in plants. It is composed of one D-glucose molecule and one D-fructose molecule. The systematic name for sucrose, *O*-α-D-glucopyranosyl-(1→2)-D-fructofuranoside, indicates four things:

- Its monosaccharides: glucose and fructose
- Their ring types: glucose is a pyranose, and fructose is a furanose
- How they are linked together: the oxygen on carbon number 1 (C1) of α-D-glucose is linked to the C2 of D-fructose.
- The *-oside* suffix indicates that the anomeric carbon of both monosaccharides participates in the glycosidic bond.

Lactose, a disaccharide composed of one D-galactose molecule and one D-glucose molecule, occurs naturally in mammalian milk. The systematic name for lactose is *O*-β-D-galactopyranosyl-(1→4)-D-glucopyranose. Other notable disaccharides include maltose (two D-glucoses linked α-1,4) and cellulobiose (two D-glucoses linked β-1,4). disaccharides can be classified into two types they are reducing and non-reducing disaccahrides if the functional group is present in bonding with another sugar unit it is called as reducing disaccharide.

Oligosaccharides and polysaccharides

Oligosaccharides and polysaccharides are composed of longer chains of monosaccharide units bound together by glycosidic bonds. The distinction between the two is based upon the number of monosaccharide units present in the chain. Oligosaccharides typically contain between three and ten monosaccharide units, and polysaccharides contain greater than ten monosaccharide units. Definitions of how large a carbohydrate must be to fall into each category vary according to personal opinion. Examples of oligosaccharides include the disaccharides mentioned above, the trisaccharide raffinose and the tetrasaccharide stachyose.

Oligosaccharides are found as a common form of protein posttranslational modification. Such posttranslational modifications include the Lewis and ABO oligosaccharides responsible for blood group classifications and so of tissue incompatibilities, the alpha-Gal epitope responsible for hyperacute rejection in xenotransplantation, and O-GlcNAc modifications.

Polysaccharides represent an important class of biological polymers. Their function in living organisms is usually either structure- or storage-related. Starch (a polymer of glucose) is used as a storage polysaccharide in plants, being found in the form of both amylose and the branched amylopectin. In animals, the structurally similar glucose polymer is the more densely branched glycogen, sometimes called 'animal starch'. Glycogen's properties allow it to be metabolized more quickly, which suits the active lives of moving animals.

Cellulose and chitin are examples of structural polysaccharides. Cellulose is used in the cell walls of plants and other organisms, and is claimed to be the most abundant organic molecule on earth. It has many uses such as a significant role in the paper and textile industries, and is used as a feedstock for the production of rayon (via the viscose process), cellulose acetate, celluloid, and nitrocellulose. Chitin has a similar structure, but has nitrogen-containing side branches, increasing its strength. It is found in arthropod exoskeletons and in the cell walls of some fungi. It also has multiple uses, including surgical threads.

Other polysaccharides include callose or laminarin, chrysolaminarin, xylan, arabinoxylan, mannan, fucoidan and galactomannan.

Carbohydrates are the main energy source for the human body. Chemically, carbohydrates are organic molecules in which carbon, hydrogen, and oxygen bond together in the ratio: $C_x(H_2O)_y$, where x and y are whole numbers that differ depending on the specific carbohydrate to which we are referring. Animals (including humans) break down carbohydrates during the process of metabolism to release energy. For example, the chemical metabolism of the sugar glucose is shown below:

$$C_6H_{12}O_6 + 6\ O_2\ 6\ CO_2 + 6\ H_2O + energy$$

Animals obtain carbohydrates by eating foods that contain them, for example potatoes, rice, breads, and so on. These carbohydrates are manufactured by plants during the process of photosynthesis. Plants harvest energy from sunlight to run the reaction just described in reverse:

$$6\ CO_2 + 6\ H_2O + energy\ (from\ sunlight)\ C_6H_{12}O_6 + 6\ O_2$$

A potato, for example, is primarily a chemical storage system containing glucose molecules manufactured during photosynthesis. In a potato, however, those glucose molecules are bound together in a long chain. As it turns out, there are two types of carbohydrates, the simple sugars and those carbohydrates that are made of long chains of sugars- the complex carbohydrates.

Simple Sugars

All carbohydrates are made up of units of sugar (also called saccharide units). Carbohydrates that contain only one sugar unit (monosaccharides) or two sugar units (disaccharides) are referred to as simple sugars. Simple sugars are sweet in taste and are broken down quickly in the body to release energy.

Two of the most common monosaccharides are glucose and fructose. Glucose is the primary form of sugar stored in the human body for energy. Fructose is the main sugar found in most fruits.

Both glucose and fructose have the same chemical formula ($C_6H_{12}O_6$); however, they have different structures, as shown (note: the carbon atoms that sit in the "corners" of the rings are not labelled):

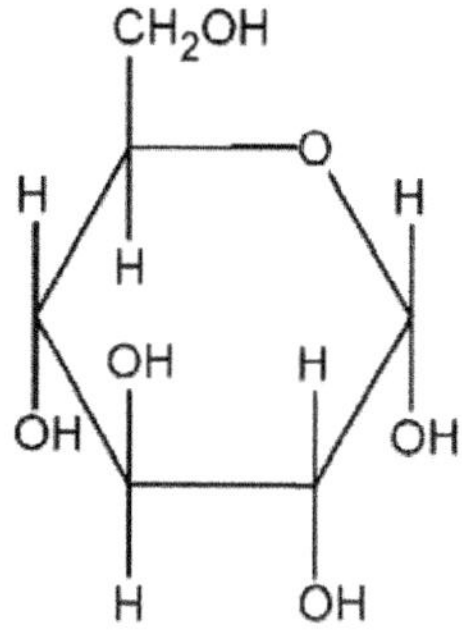

Glucose 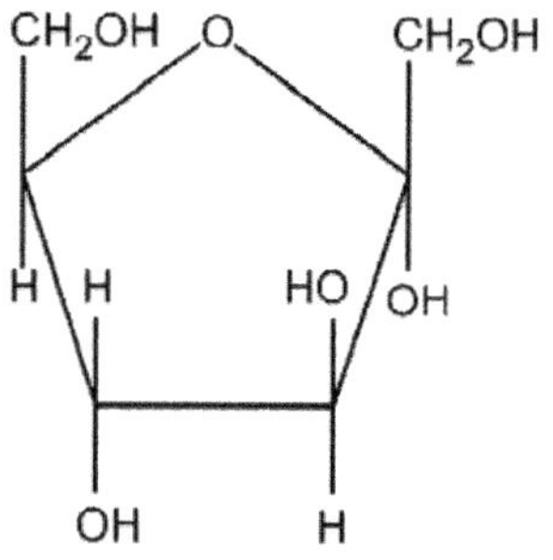   **Fructose**

Disaccharides have two sugar units bonded together. For example, common table sugar is sucrose, a disaccharide that consists of a glucose unit bonded to a fructose unit:

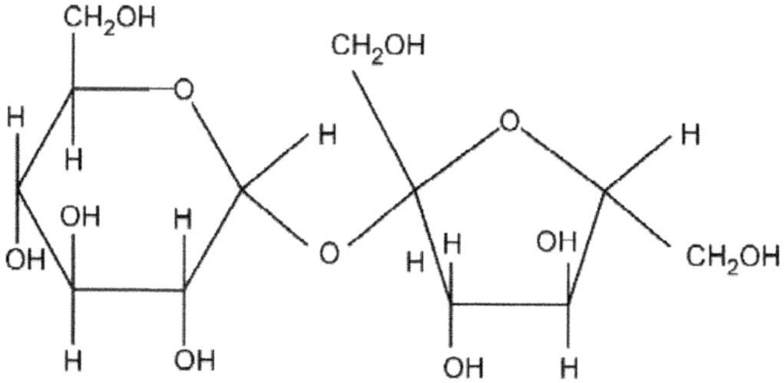

Sucrose

Complex carbohydrates Complex carbohydrates are polymers of the simple sugars. In other words, the complex carbohydrates are long chains of simple sugar units bonded together (for this reason the complex carbohydrates are often referred to as polysaccharides).

The potato we discussed earlier actually contains the complex carbohydrate starch. Starch is a polymer of the monosaccharide glucose:

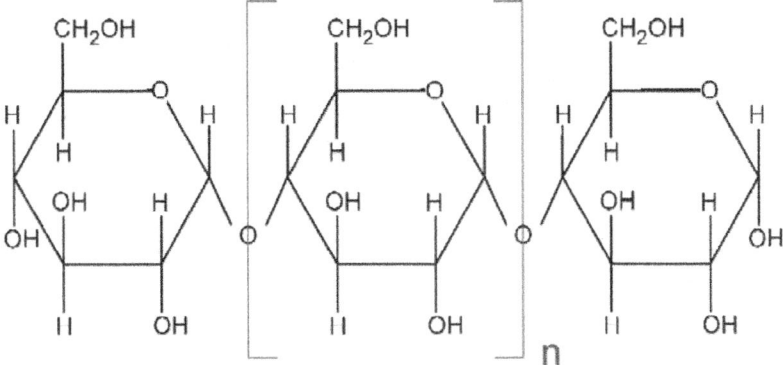

Starch (n is the number of repeating glucose units and ranges in the 1,000's)

Starch is the principal polysaccharide used by plants to store glucose for later use as energy. Plants often store starch in seeds or other specialized organs; for example, common sources of starch include rice, beans, wheat, corn, potatoes, and so on. When humans eat starch, an enzyme that occurs in saliva and in the intestines called amylase breaks the bonds between the repeating glucose units, thus allowing the sugar to be absorbed into the bloodstream.

Once absorbed into the bloodstream, the human body distributes glucose to the areas where it is needed for energy or stores it as its own special polymer-glycogen. Glycogen, another polymer of glucose, is the polysaccharide used by animals to store energy. Excess glucose is bonded together to form glycogen molecules, which the animal stores in the liver and muscle tissue as an "instant" source of energy. Both starch and glycogen are polymers of glucose; however, starch is a long, straight chain of glucose units, whereas glycogen is a branched chain of glucose units.

Another important polysaccharide is cellulose. Cellulose is yet a third polymer of the monosaccharide glucose. Cellulose differs from starch and glycogen because the glucose units form a two-dimensional structure, with hydrogen bonds holding together nearby polymers, thus giving the molecule added stability. Cellulose, also known as plant fiber, cannot be digested by human beings, therefore cellulose passes through the digestive tract without being absorbed into the body. Some animals, such as cows and termites, contain bacteria in their digestive tract that help them to digest cellulose. Cellulose is a relatively stiff material, and in plants it is used as a structural molecule to add support to the leaves, stem, and other plant parts.

Despite the fact that it cannot be used as an energy source in most animals, cellulose fiber is essential in the diet because it helps exercise the digestive track and keep it clean and healthy.

How Organisms Convert Food to Energy

Do you ever wonder how organisms convert food to energy?

We all know that organisms must take in nutrients to survive. This self-instructional unit describes two of the most important processes in food to energy conversion. These processes, called glycolysis and respiration, use the energy of the glucose molecule to manufacture ATP, the most widely used energy molecule of the cell.

ATP is an abbreviation for the nucleotide Adenosine triphosphate. This nucleotide consists of a ribose, a phosphate and the nitrogenous base adenine. This is the same nucleotide used to manufacture RNA. Don't be surprised by the fact that ATP is very energetic. All of the nucleotides used to make RNA and DNA are full of energy. Most of their energy is stored in the two bonds between their two terminal phosphates. The nucleotides of DNA are called deoxyribonucleotides (abbreviated dATP, dTTP, dCTP and dGTP). The nucleotides of RNA are called ribonucleotides (abbreviated ATP, UTP, CTP, and GTP). While these nucleotides have different types of ribose sugar and nitrogenous bases, they all have the two terminal phosphate groups, and so they all have equivalent energies. All the energy in the nucleotide comes from the two terminal phosphates.

The energy used to make DNA and RNA polymers, through the formation of phosphodiester bonds, is supplied by the triphosphate nucleotides when they polymerize to form nucleic acids. It takes energy to form all chemical bonds, and the breaking of these chemical bonds will release energy. The triphosphate nucleotides are special because the bonds between the phosphates take more energy to form (~8 kilocalories/mole) as compared to the bonds between the phosphate and the ribose (~5 kilocalories/mole). Therefore these high energy bonds between the phosphates release more energy when they are broken.

The phosphates in the structural formula of ATP have been assigned specific labels based on the IUPAC system. The phosphate attached to the #5 carbon of the ribose is labelled as the alpha (a) phosphate. The next two phosphates are labelled beta (ß) and gamma (g) from left to right. The bonds between the beta and gamma phosphates are represented by wavy lines which indicate the higher energy potential of these bonds.

Figure: The structure of ATP

Most biochemical reactions require energy input to proceed. These types of reactions are called endergonic reactions. An exergonic reaction, by contrast, gives off energy and proceeds spontaneously. Many chemical reactions of the cell use the energy from ATP which released when the beta or gamma phosphate bonds of ATP are broken. The energy released from the ATP supplies the energy necessary to form or break chemical bonds in biochemical reactions.

The energy released by the conversion of ATP to ADP is used in the manufacture of ATP during the process of glycolysis. When the phosphates are cleaved from ATP, two other forms of the nucleotide are made; ADP (adenosine diphosphate) and AMP (adenosine monophosphate). If you cover the terminal (gamma) phosphate in the structural formula of ATP with a piece of paper you will see what ADP looks like and if you cover the gamma and beta phosphates you will see what AMP looks like.

ATP is made inside cells by the transfer of a phosphate on to the b phosphate of ADP. Where does ADP come from? ADP is always inside cells as the byproduct of the cleavage of a gamma phosphate from ATP. This is the ultimate use of recycling by the cell! Sometimes the ATP, ADP and AMP molecules are completely degraded because they wear out. Cells contain many biochemical pathways and can rebuild new AMP molecules from scratch when necessary. The AMP molecule can then be phosphorylated to form ADP and ATP. However, the cell prefers to conserve energy, whenever possible, by the dephosphorylation of ATP to ADP and AMP and the rephosphorylation of these back to ATP. The following reaction represents the phosphorylation and dephosphorylation reactions of ATP.

$$ATP \rightleftharpoons ADP \rightleftharpoons AMP$$

How Cells Use Glucose for the Manufacture of ATP

The first step of the breakdown of glucose inside the cell, to utilize the potential energy stored in this molecule, is a process called glycolysis. Any basic college biology text will have all of the steps of the process of glycolysis. In this self instructional I want to give you an overview of the process, in the hope that you will understand what is taking place and not just memorize the process.

Where does glycolysis take place?

Glycolysis takes place inside the cytoplasm of the cell. The cytoplasm is the aqueous based solution inside a cell that has a variety of molecules solubilized in it. The contents of the cytoplasm include proteins, carbohydrates, nucleic acids, salts and a whole host of other soluble molecules.

What are the parts of the cell that are not considered cytoplasm?

The organelles inside the cytoplasm are not part of the cytoplasm. Remember, most organelles are bound by lipid membranes and are not soluble in the cytoplasm. The contents of organelles are also isolated from the cytoplasm due to the hydrophobic nature of their membranes. This means that the substances solubilized inside organelles stay separate from the rest of the cytoplasm. The enzymes necessary for glycolysis are solubilized in the cytoplasm of the cell.

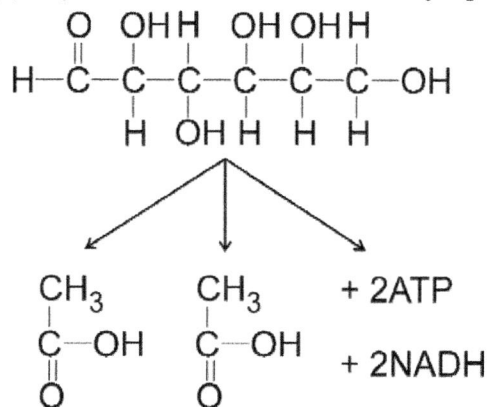

Figure: Splitting of glucose through glycolysis

What does the process of glycolysis accomplish?

* Glycolysis uses the energy from two ATP to perform a series of enzymatic reactions resulting in the manufacture of four ATP.

* Glycolysis also generates two three-carbon compounds called pyruvate that still contain energy. The destination for pyruvate

is the Citric Acid Cycle (also known as Krebs cycle) which occurs inside the matrix of the mitochondria.

- Glycolysis results in the addition of two hydrogens (electrons) onto the energy carrier nicotinamide adenine dinucleotide (NAD) to make NADH. The process of adding electrons to a substrate is called reduction. The destination for the NADH is the electron transport chain also found in the mitochondria.

Why are only two net ATP produced during glycolysis?

Although glycolysis allows for the formation of four ATP, two ATP are used in the process. This results in a net profit of two ATP from the process of Glycolysis.

Does glycolysis require molecular oxygen? No, glycolysis is an anaerobic process.

Enzymes and Enzymatic Reactions

To help understand the processes of glucose manufacture, anabolic reactions, and catabolic reactions, we need to understand how enzymes work.

Enzymes are a group of biological molecules, most of which are made of proteins. Enzymes are biological catalysts. A catalyst is added to a chemical reaction to speed the rate of the reaction. In other words, a catalyst makes the reaction occur faster. The catalyst is not changed in its form, but instead retains its original physical and chemical characteristics at the completion of the reaction. A catalyst does not participate in a reaction and can be reused again and again. Enzymes have all the qualities of a catalyst, but they are specifically used in the chemical reactions of living organisms (metabolism). Often people say; enzymes break this substance down or form this substance; however, to state the function of the enzyme correctly, one should say enzymes catalyse the formation or breakdown of a substance.

How do enzymes work? Enzymatic catalysts reduce the amount of energy necessary for endergonic reactions to proceed. Enzymes don't supply the energy, they reduce the amount needed. How? Well, there are several theories concerning this.

One way to speed reactions is to concentrate the materials that participate in them. Remember the generalized formula for chemical reactions?

Reactant A + Reactant B ———> Product C + Product D

Enzymes are molecules that have a 3-dimensional shape. Think of a muffin tin with all the depressions used to hold the individual muffins. If one hole contains reactant A and the adjacent hole contains reactant B, the two reactants are in close proximity to one another. This has the effect of concentrating the two reactants necessary for an anabolic reaction to occur and form products C and D.

Remember the cytoplasm of the cell or even the contents of an organelle may seem tiny, but it is a vast ocean of many components to a glucose or other small molecule.

Concentrating reactants is one way to speed reactions. Another way to reduce the energy and speed the rate of catabolic reactions might be to hold a molecule in place so that some of its bonds are under strain. Imagine you were trying to cut a thick piece of cloth with dull scissors and the going was very slow. If two people were to hold the material taut, or under strain, the fabric would cut more easily. Perhaps enzymes hold molecules in a similar fashion allowing bonds to be broken more easily.

If enzymes only speed the rates of reactions, does this mean that chemical reactions occur spontaneously anyway? Yes, all chemical reactions can occur spontaneously, but some reactions may take millennia to occur. Since organisms do not live on a geological time scale, the ability of enzymes to speed the rate of reactions, up to thousands of times faster, is necessary for life to exist.

The Structure of Enzymes

Enzymes are 3-dimensional molecules that have folded up into a specific shape (eg. the muffin tin). Imagine the depression in our muffin tin to be the site at which the chemical reaction takes place. This site is called the active site of the enzyme. The molecule that fits in the active site of the enzyme is called the substrate. Enzymes and substrates are very specific for each other. The active site of an enzyme is designed to fit perfectly with its specific substrate, like a 3-dimensional jigsaw puzzle.

Biochemical or Enzymatic Pathways

Sometimes a cell needs a product from a reactant that requires only a single chemical reaction. For example, reactant A needs to be changed to product B. This reaction would only require the enzyme that fits with its substrate, reactant A. However, this is not usually the case. Most products inside cells that need to be broken down or built up require a multi-step process. This process is called a biochemical pathway. It goes something like this:

- Enzyme A changes substrate A to product B.
- Product or substrate B now can be bound by enzyme B and be changed to product or substrate C
- Substrate C now fits with enzyme C which changes it to product D.
- If product D is what the cell needs, the pathway is finished. If not, the pathway can continue in the same way until the ultimate product is obtained.

Finally a quick lesson in naming enzymes. Generally to name an enzyme we take the substrate it breaks down or product that it manufactures and add an -ase on the end.

Examples: enzymes that degrade the following substances DNAse, ATPase or enzymes that build up the following substances: DNA polymerase and ATP synthetase.

Glucose Utilization

The figure below demonstrates some of the possible pathways for glucose utilization by a variety of organisms.

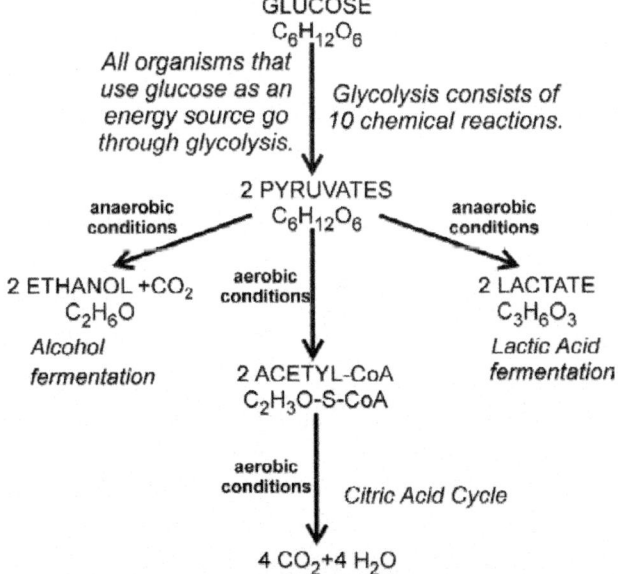

GLUCOSE
$C_6H_{12}O_6$

All organisms that use glucose as an energy source go through glycolysis.

Glycolysis consists of 10 chemical reactions.

2 PYRUVATES
$C_6H_{12}O_6$

anaerobic conditions

anaerobic conditions

2 ETHANOL $+CO_2$
C_2H_6O

aerobic conditions

2 LACTATE
$C_3H_6O_3$

Alcohol fermentation

Lactic Acid fermentation

2 ACETYL-CoA
$C_2H_3O\text{-}S\text{-}CoA$

aerobic conditions

Citric Acid Cycle

4 CO_2+4 H_2O

Figure: Pyruvate metabolism in different organisms

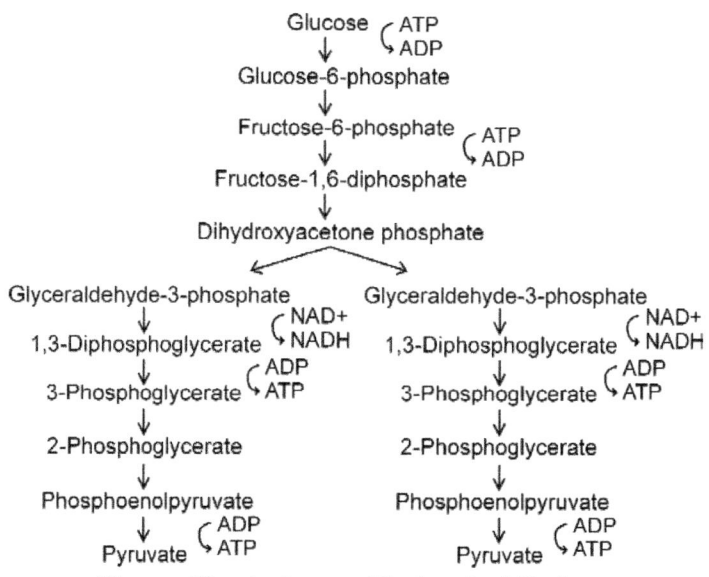

Figure: Glycolysis as a Biochemical Pathway

Please note that the conversion of glucose to pyruvate requires 10 enzyme reactions and occurs in 10 separate steps. After glycolysis, what happens to the pyruvate molecules?

Both of the pyruvate (three-carbon molecules) are immediately converted to two-carbon molecules by removing a CO_2 (carbon dioxide) from each molecule. The remaining two-carbon compound is known as an acetyl group. The acetyl group will react quickly with another molecule because of its unpaired electrons. Therefore as the CO_2 is cleaved from pyruvate, the acetyl group is added simultaneously to a carrier molecule called coenzyme A (CoA).

Coenzyme A (CoA) is not an enzyme but a molecule similar in some respects to a nucleotide. This is a very important carrier molecule because it carries the acetyl group to many important biochemical pathways. The acetyl group and CoA molecule combine to form acetyl CoA. This molecule travels to the matrix of the mitochondria.

Once there, the CoA transfers the two-carbon acetyl group to an enzyme which catalyzes the addition of a four-carbon compound called oxaloacetic acid (oxaloacetate) to the acetyl. The result of this reaction is a six-carbon compound called citric acid (citrate). This is why this biochemical pathway is called the citric acid cycle.

The citric acid molecule has carbon, hydrogen, and oxygen atoms stripped from it in a step-wise fashion through this biochemical pathway. Each step requires its own specific enzyme. The end result of this pathway is the regeneration of oxaloacetic acid, and the conversion of NAD and FAD to NADH and FADH with the formation of carbon dioxide CO_2 as a byproduct.

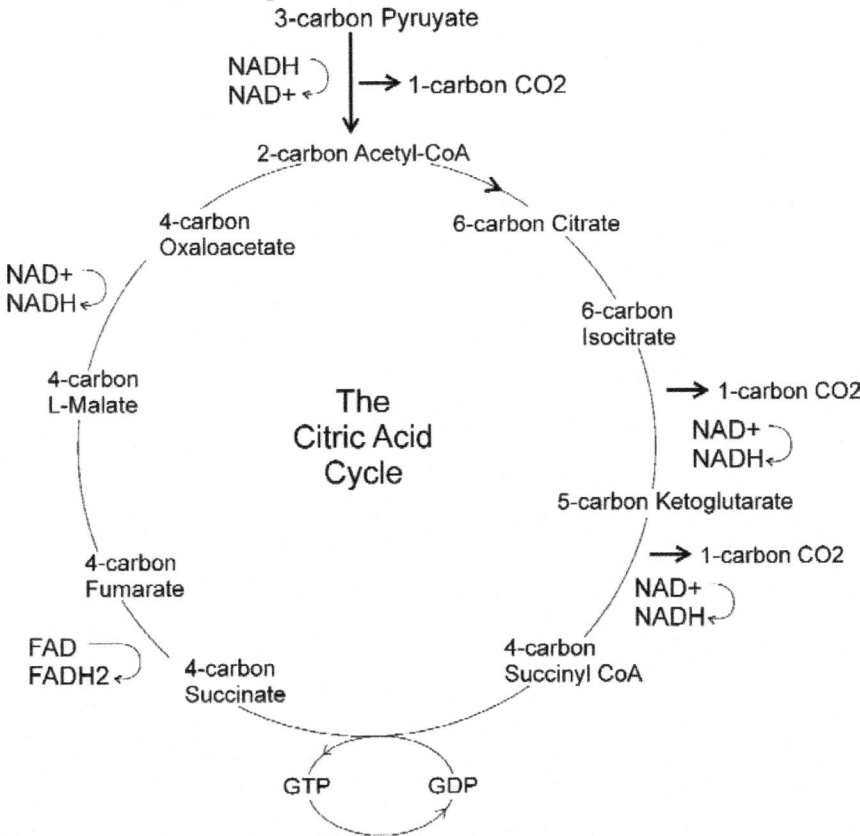

Figure: A simplified diagram of the citric acid (Krebs) cycle.

The major objective of the citric acid cycle is to get the hydrogens, along with their electrons, off the carbon compounds so they can be carried to the electron transport proteins that are embedded in the membrane of the mitochondrion.

The waste or by-products of these reactions are the carbon and oxygen that are stripped off as each compound is rearranged by the removal of hydrogen. These carbons and oxygens combine to form CO_2 carbon dioxide gas which is given off as a byproduct of these reactions. The CO_2 that you breathe out of your lungs is a waste product coming from the cellular metabolism of glucose! Carbon dioxide (CO_2) enters our bloodstream, eventually winds up in our lungs, and is expelled upon exhalation, but where do those hydrogens with their electrons go? The hydrogens and electrons are picked up by electron carrier NAD (nicotinamide adenine dinucleotide) and electron carrier FAD (flavin adenine dinucleotide).

These electron carriers transfer the hydrogens and electrons to the electron carrying proteins, embedded in the mitochondrial membrane, which form the electron transport chain. The electron transport chain proteins are complexed with metal ions which allow them to carry extra electrons safely.

When NADH and FADH arrive at the electron transport chain, they transfer the electrons associated with the hydrogens to the electron carrying proteins at the top of the electron chain.

Figure: The structural formula for NAD and NADH.

When electrons and hydrogen atoms are added to a molecule, this is called reduction. When electrons or hydrogens are removed from a molecule, this is known as oxidation. NAD is in an oxidized state while NADH is in a reduced state.

Now that the electrons have been transferred to the electron transport proteins, the hydrogen ion $H+$ (proton) is free to be pumped across the membrane into a space between the inner and outer mitochondrial membranes.

This proton pumping requires energy which is generated by the flow of electrons as they are passed down the electron transport protein chain. As electrons are transferred from one protein to the next, energy is released at each transfer, until the final electron acceptor, oxygen is reached. Aerobic organisms must take in oxygen because it serves as the final electron acceptor in the electron transport chain. The energy, released from the passage of the electrons down the electron transport chain to oxygen, is used to pump the protons across the membrane. The pumping of protons requires a great deal of energy, because they have to be pumped against a concentration gradient.

Because protons are constantly being pumped across the membrane, they fill the space between the mitochondrial membranes. This is known as an area of high proton concentration. Putting more protons in this area would be like trying to force balloons into a box already filled to capacity with balloons.

Pushing protons against a concentration gradient of protons is referred to as going against, or up, a concentration gradient. In the case of the balloons, as soon as you open the box they would come out easily with no energy input necessary.

This is termed going with, or down, the concentration gradient. Molecules naturally flow from areas of high concentration to areas of low concentration with no energy input necessary.

Once you have captured all this energy by pushing protons up the proton concentration gradient into a crowded area, what happens? The protons can get out but they only have one passage to escape. This is a passage created by a protein enzyme called ATP synthetase. The ATP synthetase phosphorylates (adds phosphates to) ADP. The ATP synthetase gets the energy necessary for the formation of the high energy phosphate bond from the rush of protons through the channel it creates across the mitochondrial membrane.

To better understand how the flow of protons provides energy for the ATP synthetase enzyme imagine how a turbine engine works at Hoover Dam. The rushing water is caught and causes the huge turbine to turn which generates electricity. In a similar way the energy of protons cascading through the ATP synthetase is converted into the energy used to form the phosphate bond of ATP.

This process, called oxidative phosphorylation, makes most of the ATP manufactured by cells. While glycolysis gives us net profit of two ATP, the electron transport coupled with the citric acid cycle and oxidative phosphorylation give us a net profit of approximately 34 ATP from each glucose catabolized. Quite a big difference in yield!

What happens to the protons and electrons when they flow back through the membrane and down electron transport chain? Since electrons are energetic enough to cause problems inside the cell they must be inactivated. The final electron acceptor, oxygen, combines with the electrons and excess hydrogen ions to form water, a non-toxic by-product.

So the products of respiration are CO_2 (from the citric acid cycle) and H_2O (from electron transport) and lots of ATP. Without oxygen, aerobic cells die because the manufacture of ATP is halted at the cellular level. Since oxygen is absolutely necessary for the continual phosphorylation of ADP to ATP this process is called oxidative phosphorylation. Without oxygen, electron transport would stop, protons would not be pumped, and the ATP synthetase would not have a flow of protons to supply the energy necessary for the manufacture of ATP.

Do you know where all the molecules come from and their destination in the following Generalized Equation for Respiration?

$$C6H_{12}O_6 + O_2 \longrightarrow CO_2 + H_2O + energy$$

Now I would like to give you my overview of all of the processes of glucose metabolism in narrative form. I hope that you will understand the processes of glycolysis and respiration well enough to be able to describe this process in a similar way. In other words, I would like you to be able to think through and understand the concepts, not the minute details of these processes.

- Glucose is taken up as food by the body and carried to a cell. Glucose enters the cell by facilitated diffusion and is immediately broken down by the process of glycolysis in the cytoplasm of the cell.

- The process of glycolysis uses a biochemical pathway consisting of 10 enzyme mediated reactions which uses two ATP and manufactures four ATP (a net profit of two ATP). Glycolysis generates two NADH, which carry hydrogen and electrons to the electron transport chain in the mitochondrial membrane. Glycolysis also generates two three-carbon pyruvate molecules.

- The pyruvate molecule has a CO_2 removed and the remaining two-carbon acetyl group is then added onto the carrier molecule Coenzyme A to form acetyl CoA.

- Acetyl CoA is taken to the mitochondrial matrix to participate in the Citric acid cycle. The two-carbon acetyl group is then added via an enzyme mediated reaction to a four-carbon oxaloacetic acid molecule which results in the formation of the six-carbon molecule citric acid. Through a series of enzyme mediated reactions oxygens, carbons and hydrogens are systematically stripped off. The carbons and oxygens combine in the appropriate proportions to form CO2 and ultimately are expelled as waste products.

- The hydrogen atoms along with their electrons are picked up by the electron carriers NAD and FAD and taken to the electron transport proteins, which are located on the inner mitochondrial membrane.

- The electrons are handed off to the electron transport proteins where they flow from one protein to another releasing energy at each transfer until they combine with oxygen, the final electron acceptor in the electron transport chain.

- The energy released from the flow of electrons down the transport chain is used to pump protons across the inner mitochondrial membrane into the space between the inner and outer mitochondrial membranes, against a proton concentration gradient.

- The protons then flow back down the concentration gradient through an ATP synthetase enzyme channel, back into the mitochondrial matrix. The flow of protons through the ATP synthetase generates the energy necessary to phosphorylate ADP to ATP.

- Once back inside the matrix the protons combine with oxygen and electrons to form water a by-product of oxidative phosphorylation.

Protein Structure and Function

Proteins (also known as polypeptides) are organic compounds made of amino acids arranged in a linear chain and folded into a globular form. The amino acids in a polymer are joined together by the peptide bonds between the carboxyl and amino groups of adjacent amino acid residues. The sequence of amino acids in a protein is defined by the sequence of a gene, which is encoded in the genetic code. In general, the genetic code specifies 20 standard amino acids; however, in certain organisms the genetic code can include selenocysteine—and in certain archaea—pyrrolysine. Shortly after or even during synthesis, the residues in a protein are often chemically modified by post-translational modification, which alters the physical and chemical properties, folding, stability, activity, and ultimately, the function of the proteins. Proteins can also work together to achieve a particular function, and they often associate to form stable complexes.

Of the most distinguishing features of polypeptides is their ability to fold into a globular state, or "structure". The extent to which proteins fold into a defined structure varies widely. Data supports that some protein structures fold into a highly rigid structure with small fluctuations and are therefore considered to be single structure. Other proteins have been shown to undergo large rearrangements from one conformation to another. This conformational change is often associated with a signalling event. Thus, the structure of a protein serves as a medium through which to regulate either the function of a protein or activity of an enzyme. Not all proteins requiring a folding process in order to function as some function in an unfolded state.

Like other biological macromolecules such as polysaccharides and nucleic acids, proteins are essential parts of organisms and participate in virtually every process within cells. Many proteins are enzymes that catalyse biochemical reactions and are vital to metabolism. Proteins also have structural or mechanical functions, such as actin and myosin in muscle and the proteins in the cytoskeleton, which form a system of scaffolding that maintains cell shape. Other proteins are important in cell signalling, immune responses, cell adhesion, and the cell cycle. Proteins are also necessary in animals' diets, since animals cannot synthesize all the amino acids they need and must obtain essential amino acids from food. Through the process of digestion, animals break down ingested protein into free amino acids that are then used in metabolism.

Proteins were first described by the Dutch chemist Gerhardus Johannes Mulder and named by the Swedish chemist Jöns Jakob Berzelius in 1838. Early nutritional scientists such as the German Carl von Voit believed that protein was the most important nutrient for maintaining the structure of the body, because it was generally believed that "flesh makes flesh." The central role of proteins as enzymes in living organisms was however not fully appreciated until 1926, when James B. Sumner showed that the enzyme urease was in fact a protein. The first protein to be sequenced was insulin, by Frederick Sanger, who won the Nobel Prize for this achievement in 1958. The first protein structures to be solved were hemoglobin and myoglobin, by Max Perutz and Sir John Cowdery Kendrew, respectively, in 1958. The three-dimensional structures of both proteins were first determined by x-ray diffraction analysis; Perutz and Kendrew shared the 1962 Nobel Prize in Chemistry for these discoveries. Proteins may be purified from other cellular components using a variety of techniques such as ultracentrifugation, precipitation, electrophoresis, and chromatography; the advent of genetic engineering has made possible a number of methods to facilitate purification. Methods commonly used to study protein structure and function include immunohistochemistry, site-directed mutagenesis, nuclear magnetic resonance and mass spectrometry.

Most proteins are linear polymers built from series of up to 20 different L-α-amino acids. All amino acids possess common structural features, including an α-carbon to which an amino group, a carboxyl group, and a variable side chain are bonded. Only proline differs from this basic structure as it contains an unusual ring to the N-end amine group, which forces the CO–NH amide moiety into a fixed conformation. The side chains of the standard amino acids, detailed in the list of standard amino acids, have a great variety of chemical structures and properties; it is the combined effect of all of the amino acid side chains in a protein that ultimately determines its three-dimensional structure and its chemical reactivity.

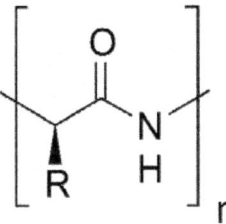

Chemical structure of the peptide bond (left) and a peptide bond between leucine and threonine (right).

The amino acids in a polypeptide chain are linked by peptide bonds. Once linked in the protein chain, an individual amino acid is called a *residue,* and the linked series of carbon, nitrogen, and oxygen atoms are known as the *main chain* or *protein backbone*. The peptide bond has two resonance forms that contribute some double-bond character and inhibit rotation around its axis, so that the alpha carbons are roughly coplanar. The other two dihedral angles in the peptide bond determine the local shape assumed by the protein backbone. The end of the protein with a free carboxyl group is known as the C-terminus or carboxy terminus, whereas the end with a free amino group is known as the N-terminus or amino terminus.

The words *protein, polypeptide,* and *peptide* are a little ambiguous and can overlap in meaning. *Protein* is generally used to refer to the complete biological molecule in a stable conformation, whereas *peptide* is generally reserved for a short amino acid oligomers often lacking a stable three-dimensional structure. However, the boundary between the two is not well defined and usually lies near 20–30 residues. *Polypeptide* can refer to any single linear chain of amino acids, usually regardless of length, but often implies an absence of a defined conformation.

Synthesis

Proteins are assembled from amino acids using information encoded in genes. Each protein has its own unique amino acid sequence that is specified by the nucleotide sequence of the gene encoding this protein. The genetic code is a set of three-nucleotide sets called codons and each three-nucleotide combination designates an amino acid, for example AUG (adenine-uracil-guanine) is the code for methionine. Because DNA contains four nucleotides, the total number of possible codons is 64; hence, there is some redundancy in the genetic code, with some amino acids specified by more than one codon.

Genes encoded in DNA are first transcribed into pre-messenger RNA (mRNA) by proteins such as RNA polymerase. Most organisms then process the pre-mRNA (also known as a *primary transcript*) using various forms of post-transcriptional modification to form the mature mRNA, which is then used as a template for protein synthesis by the ribosome. In prokaryotes the mRNA may either be used as soon as it is produced, or be bound by a ribosome after having moved away from the nucleoid. In contrast, eukaryotes make mRNA in the cell nucleus and then translocate it across the nuclear membrane into the cytoplasm, where protein synthesis then takes place. The rate of protein synthesis is higher in prokaryotes than eukaryotes and can reach up to 20 amino acids per second.

The process of synthesizing a protein from an mRNA template is known as translation. The mRNA is loaded onto the ribosome and is read three nucleotides at a time by matching each codon to its base pairing anticodon located on a transfer RNA molecule, which carries the amino acid corresponding to the codon it recognizes. The enzyme aminoacyl tRNA synthetase "charges" the tRNA molecules with the correct amino acids. The growing polypeptide is often termed the *nascent chain*. Proteins are always biosynthesized from N-terminus to C-terminus. The size of a synthesized protein can be measured by the number of amino acids it contains and by its total molecular mass, which is normally reported in units of *daltons* (synonymous with atomic mass units), or the derivative unit kilodalton (kDa). Yeast proteins are on average 466 amino acids long and 53 kDa in mass. The largest known proteins are the titins, a component of the muscle sarcomere, with a molecular mass of almost 3,000 kDa and a total length of almost 27,000 amino acids.

Chemical Synthesis

Short proteins can also be synthesized chemically by a family of methods known as peptide synthesis, which rely on organic synthesis techniques such as chemical ligation to produce peptides in high yield. Chemical synthesis allows for the introduction of non-natural amino acids into polypeptide chains, such as attachment of fluorescent probes to amino acid side chains.

These methods are useful in laboratory biochemistry and cell biology, though generally not for commercial applications. Chemical synthesis is inefficient for polypeptides longer than about 300 amino acids, and the synthesized proteins may not readily assume their native tertiary structure. Most chemical synthesis methods proceed from C-terminus to N-terminus, opposite the biological reaction.

Structure

Proteins are an important class of biological macromolecules present in all organisms. All proteins are polymers of amino acids. Classified by their physical size, proteins are nanoparticles (definition: 1-100 nm). Each protein polymer – also known as a polypeptide – consists of a sequence of 20 different L-α-amino acids, also referred to as residues. For chains under 40 residues the term peptide is frequently used instead of protein.

To be able to perform their biological function, proteins fold into one or more specific spatial conformations, driven by a number of non-covalent interactions such as hydrogen bonding, ionic interactions, Van Der Waals forces, and hydrophobic packing. To understand the functions of proteins at a molecular level, it is often necessary to determine their three-dimensional structure. This is the topic of the scientific field of structural biology, which employs techniques such as X-ray crystallography, NMR spectroscopy, and dual polarisation interferometry to determine the structure of proteins.

Protein structures range in size from tens to several thousand residues Very large aggregates can be formed from protein subunits: for example, many thousand actin molecules assemble into a microfilament.

A protein may undergo reversible structural changes in performing its biological function. The alternative structures of the same protein are referred to as different conformations, and transitions between them are called conformational changes.

Protein Covalent Structure and Stereochemistry

Protein amino acids are combined into a single polypeptide chain in a condensation reaction. This reaction is catalysed by the ribosome in a process known as translation.

Amino Acid Residues

Each α-amino acid consists of a backbone part that is present in all the amino acid types, and a side chain that is unique to each type of residue. An exception from this rule is proline, where the hydrogen atom is replaced by a bond to the side chain.

Because the carbon atom is bound to four different groups it is chiral, however only one of the isomers occur in biological proteins. Glycine however, is not chiral since its side chain is a hydrogen atom. A simple mnemonic for correct L-form is "CORN": when the C_α atom is viewed with the H in front, the residues read "CO-R-N" in a clockwise direction.

The 20 naturally occurring amino acids have different physical and chemical properties, including their electrostatic charge, pKa, hydrophobicity, size and specific functional groups. These properties play a major role in molding protein structure.

The Peptide Bond

The peptide bond tend to be planar due to the delocalization of the electrons from the double bond. The rigid peptide dihedral angle, ω (the bond between C_1 and N) is always close to 180 degrees. The dihedral angles phi φ (the bond between N and Cα) and psi $\o$ (the bond between Cα and C_1) can have a certain range of possible values.

These angles are the internal degrees of freedom of a protein, they control the protein's conformation. They are restrained by geometry to allowed ranges typical for particular secondary structure elements, and represented in a Ramachandran plot.

Side-chain Conformation

The atoms along the side chain are named with Greek letters in Greek alphabetical order: α, β, γ, δ, °, and so on. C_α refers to the carbon atom of the backbone closest to the carbonyl group of that amino acid, C_β the second closest and so on. The dihedral angles around the bonds between these atoms are named ÷1, ÷2, ÷3, etc. The dihedral angle of the first movable atom of the side chain, γ, defined as N-Cα-Cβ-Xγ, is named ÷1. Side chains tend to adopt different staggered conformations called *gauche(-)*, *trans*, and *gauche(+)*, which corresponds to rotation angles of 60°, 180°, and -60°, respectively, around the sp3-sp3 bonds.

The diversity of side-chain conformations is often expressed in rotamer libraries. A rotamer library is a collection of rotamers for each residue type. Side-chain dihedral angles are not evenly distributed, but for most side chain types, the ÷ angles occur in tight clusters around certain values. Rotamer libraries therefore are usually derived from statistical analysis of side-chain conformations in known structures of proteins by clustering observed conformations or by dividing dihedral angle space into bins, and determining an average conformation in each bin.

Levels of Protein Structure

There are four distinct levels of protein structure.

Primary Structure

The primary structure refers to the sequence of the different amino acids of the peptide or protein. The primary structure is held together by covalent or peptide bonds, which are made during the process of protein biosynthesis or translation. The two ends of the polypeptide chain are referred to as the carboxyl terminus (C-terminus) and the amino terminus (N-terminus) based on the nature of the free group on each extremity. Counting of residues always starts at the N-terminal end (NH_2-group), which is the end where the amino group is involved in a peptide bond.

The primary structure of a protein is determined by the gene corresponding to the protein. A specific sequence of nucleotides in DNA is transcribed into mRNA, which is read by the ribosome in a process called translation. The sequence of a protein is unique to that protein, and defines the structure and function of the protein. The sequence of a protein can be determined by methods such as Edman degradation or tandem mass spectrometry. Often however, it is read directly from the sequence of the gene using the genetic code. Post-translational modifications such as disulfide formation, phosphorylations and glycosylations are usually also considered a part of the primary structure, and cannot be read from the gene.

Secondary Structure

Secondary structure can be formally defined by the hydrogen bonds of the biopolymer, as observed in an atomic-resolution structure. In proteins, the secondary structure is defined by the patterns of hydrogen bonds between backbone amide and carboxyl groups. In nucleic acids, the secondary structure is defined by the hydrogen bonding between the nitrogenous bases. The hydrogen bonding patterns may be significantly distorted, which makes an automatic determination of secondary structure difficult.

The secondary structure may be also defined based on the regular pattern of backbone dihedral angles in a particular region of the Ramachandran plot; thus, a segment of residues with such dihedral angles may be called a helix, regardless of whether it has the correct hydrogen bonds. The secondary structure may be also provided by crystallographers in the corresponding PDB file.

The rough secondary-structure content of a biopolymer (e.g., "this protein is 40% α-helix and 20% β-sheet.") can often be estimated spectroscopically. For proteins, a common method is far-ultraviolet (far-UV, 170-250 nm) circular dichroism. A pronounced double minimum at 208 and 222 nm indicate α-helical structure, whereas a single minimum at 204 nm or 217 nm reflects random-coil or β-sheet structure, respectively. A less common method is infrared spectroscopy, which detects differences in the bond oscillations of amide groups due to hydrogen-bonding. Finally, secondary-structure contents may be estimated accurately using the chemical shifts of an unassigned NMR spectrum.

Protein

Secondary structure in proteins consists of local inter-residue interactions mediated by hydrogen bonds, or not. The most common secondary structures are alpha helices and beta sheets. Other helices, such as the 3_{10} helix and ð helix, are calculated to have energetically favourable hydrogen-bonding patterns but are rarely if ever observed in natural proteins except at the ends of α helices due to unfavourable backbone packing in the centre of the helix. Other extended structures such as the polyproline helix and alpha sheet are rare in native state proteins but are often hypothesized as important protein folding intermediates. Tight turns and loose, flexible loops link the more "regular" secondary structure elements. The random coil is not a true secondary structure, but is the class of conformations that indicate an absence of regular secondary structure.

Amino acids vary in their ability to form the various secondary structure elements. Proline and glycine are sometimes known as "helix breakers" because they disrupt the regularity of the α helical backbone conformation; however, both have unusual conformational abilities and are commonly found in turns. Amino acids that prefer to adopt helical conformations in proteins include methionine, alanine, leucine, glutamate and lysine ("MALEK" in amino-acid 1-letter codes); by contrast, the large aromatic residues (tryptophan, tyrosine and phenylalanine) and C^β-branched amino acids (isoleucine, valine, and threonine) prefer to adopt β-strand conformations. However, these preferences are not strong enough to produce a reliable method of predicting secondary structure from sequence alone.

There are several methods for defining protein secondary structure (e.g. DEFINE, DSSP, STRIDE (protein)).

The DSSP Code

The Dictionary of Protein Secondary Structure, in short DSSP, is commonly used to describe the protein secondary structure with single letter codes. The secondary structure is assigned based on hydrogen bonding patterns as those initially proposed by Pauling et al. in 1951 (before any protein structure had ever been experimentally determined). There are eight types of secondary structure that DSSP defines:

- G = 3-turn helix (3_{10} helix). Min length 3 residues.
- H = 4-turn helix (α helix). Min length 4 residues.
- I = 5-turn helix (ð helix). Min length 5 residues.
- T = hydrogen bonded turn (3, 4 or 5 turn)
- E = extended strand in parallel and/or anti-parallel b-sheet conformation. Min length 2 residues.
- B = residue in isolated b-bridge (single pair b-sheet hydrogen bond formation)
- S = bend (the only non-hydrogen-bond based assignment).

Amino acid residues which are not in any of the above conformations are assigned as the eighth type 'Coil': often codified as ' ' (space), C (coil) or '-' (dash). The helices (G,H and I) and sheet conformations are all required to have a reasonable length. This means that 2 adjacent residues in the primary structure must form the same hydrogen bonding pattern. If the helix or sheet hydrogen bonding pattern is too short they are designated as T or B, respectively. Other protein secondary structure assignment categories exist (sharp turns, Omega loops etc.), but they are less frequently used.

DSSP H-bond Definition

Secondary structure is defined by hydrogen bonding, so the exact definition of a hydrogen bond is critical. The standard H-bond definition for secondary structure is that of DSSP, which is a purely electrostatic model. It assigns charges of $\pm q_1 \equiv 0.42e$ to the carbonyl carbon and oxygen, respectively, and charges of $\pm q_2 \equiv 0.20e$ to the amide nitrogen and hydrogen, respectively. The electrostatic energy is:

$$E = q_1 q_2 \left[\frac{1}{r_{ON}} + \frac{1}{r_{CH}} - \frac{1}{r_{OH}} - \frac{1}{r_{CN}} \right] .332 \, \text{kcal/mol.}$$

According to DSSP, an H-bond exists if and only if E is less than -0.5 kcal/mol. Although the DSSP formula is a relatively crude approximation of the *physical* H-bond energy, it is generally accepted as a tool for defining secondary structure.

Protein Secondary-structure Prediction

Predicting protein tertiary structure from only its amino acid sequence is a very challenging problem, but using the simpler secondary structure definitions is more tractable and has been the focus for research for a long time.

Although, the 8-state DSSP code is already a simplification from the continuous variation of hydrogen bonding patterns present in a protein the majority of secondary prediction methods simplify further to the three dominant states: Helix, Sheet and Coil. How the conversion is made from 8- to 3-state varies between methods.

Early methods of secondary-structure prediction were based on the helix- or sheet-forming propensities of individual amino acids, sometimes coupled with rules for estimating the free energy of forming secondary structure elements. Such methods were typically ~60% accurate in predicting which of the three states (helix/sheet/coil) a residue adopts. A significant increase in accuracy (to nearly ~80%) was made by exploiting multiple sequence alignment; knowing the full distribution of amino acids that occur at a position (and in its vicinity, typically ~7 residues on either side) throughout evolution provides a much better picture of the structural tendencies near that position.

For illustration, a given protein might have a glycine at a given position, which by itself might suggest a random coil there. However, multiple sequence alignment might reveal that helix-favouring amino acids occur at that position (and nearby positions) in 95% of homologous proteins spanning nearly a billion years of evolution.

Moreover, by examining the average hydrophobicity at that and nearby positions, the same alignment might also suggest a pattern of residue solvent accessibility consistent with an α-helix. Taken together, these factors would suggest that the glycine of the original protein adopts α-helical structure, rather than random coil. Several types of methods are used to combine all the available data to form a 3-state prediction, including neural networks, hidden Markov models and support vector machines. Modern prediction methods also provide a confidence score for their predictions at every position.

Secondary-structure prediction methods are continuously benchmarked, e.g., in the EVA experiment. Based on ~270 weeks of testing, the most accurate methods at present are PSIPRED, SAM, PORTER, PROF and SABLE. Interestingly, it does not seem to be possible to improve upon these methods by taking a consensus of them The chief area for improvement appears to be the prediction of β-strands; residues confidently predicted as β-strand are likely to be so, but the methods are apt to overlook some β-strand segments (false negatives). There is likely an upper limit of ~90% prediction accuracy overall, due to the idiosyncrasies of the standard method (DSSP) for assigning secondary-structure classes (helix/strand/coil) to PDB structures, against which the predictions are benchmarked

Accurate secondary-structure prediction is a key element in the prediction of tertiary structure, in all but the simplest (homology modeling) cases.

Alignment

Both protein and nucleic acid secondary structures can be used to aid in multiple sequence alignment. These alignments can be made more accurate by the inclusion of secondary structure information in addition to simple sequence information. This is sometimes less useful in RNA because base pairing is much more highly conserved than sequence. Distant relationships between proteins whose primary structures are unalignable can sometimes be found by secondary structure.

Tertiary Structure

Tertiary structure refers to three-dimensional structure of a single protein molecule. The alpha-helices and beta-sheets are folded into a compact globule. The folding is driven by the *non-specific* hydrophobic interactions (the burial of hydrophobic residues from water), but the structure is stable only when the parts of a protein domain are locked into place by *specific* tertiary interactions, such as salt bridges, hydrogen bonds, and the tight packing of side chains and disulfide bonds. The disulfide bonds are extremely rare in cytosolic proteins, since the cytosol is generally a reducing environment.

Relationship to Primary Structure

Tertiary structure is considered to be largely determined by the protein's primary structure, or the sequence of amino acids of which it is composed. Efforts to predict tertiary structure from the primary structure are known generally as protein structure prediction.

However, the environment in which a protein is synthesized and allowed to fold are significant determinants of its final shape and are usually not directly taken into account by current prediction methods. (Most such methods do rely on comparisons between the sequence to be predicted and sequences of known structure in the Protein Data Bank and thus account for environment indirectly, assuming the target and template sequences share similar cellular contexts.)

Determinants of Tertiary Structure

In globular proteins, tertiary interactions are frequently stabilized by the sequestration of hydrophobic amino acid residues in the protein core, from which water is excluded, and by the consequent enrichment of charged or hydrophilic residues on the protein's water-exposed surface. In secreted proteins that do not spend time in the cytoplasm, disulfide bonds between cysteine residues help to maintain the protein's tertiary structure. A variety of common and stable tertiary structures appear in a large number of proteins that are unrelated in both function and evolution- for example, many proteins are shaped like a TIM barrel, named for the enzyme triosephosphateisomerase. Another common structure is a highly stable dimeric coiled coil structure composed of 2-7 alpha helices. Proteins are classified by the folds they represent in databases like SCOP and CATH.

Stability of Native States

The most typical conformation of a protein in its cellular environment is generally referred to as the native state or native conformation. It is commonly assumed that this most-populated state is also the most thermodynamically stable conformation attainable for a given primary structure; this is a reasonable first approximation but the claim assumes that the reaction is not under kinetic control- that is, that the time required for the protein to attain its native conformation before being translated is small.

In the cell, a variety of protein chaperones assist a newly synthesized polypeptide in attaining its native conformation. Some such proteins are highly specific in their function, such as protein disulfide isomerase; others are very general and can be of assistance to most globular proteins- the prokaryotic GroEL/GroES system and the homologous eukaryotic Heat shock proteins Hsp60/Hsp10 system fall into this category.

Some proteins explicitly take advantage of the fact that they can become kinetically trapped in a relatively high-energy conformation due to folding kinetics. Influenza hemagglutinin, for example, is synthesized as a single polypeptide chain that acts as a kinetic trap. The "mature" activated protein is proteolytically cleaved to form two polypeptide chains that are trapped in a high-energy conformation. Upon encountering a drop in pH, the protein undergoes an energetically favourable conformational rearrangement that enables it to penetrate a host cell membrane.

Many serpins (serine protease inhibitors) are metastable, and undergo a conformational change when a loop of the protein is cut by a protease.

Quaternary Structure

Quaternary structure is a larger assembly of several protein molecules or polypeptide chains, usually called subunits in this context. The quaternary structure is stabilized by the same non-covalent interactions and disulfide bonds as the tertiary structure. Complexes of two or more polypeptides (i.e. multiple subunits) are called multimers. Specifically it would be called a dimer if it contains two subunits, a trimer if it contains three subunits, and a tetramer if it contains four subunits. The subunits are frequently related to one another by symmetry operations, such as a 2-fold axis in a dimer. Multimers made up of identical subunits are referred to with a prefix of "homo-" (e.g. a homotetramer) and those made up of different subunits are referred to with a prefix of "hetero-" (e.g. a heterotetramer, such as the two alpha and two beta chains of hemoglobin). Many proteins do not have the quaternary structure and function as monomers.

Many proteins are actually assemblies of more than one polypeptide chain, which in the context of the larger assemblage are known as protein subunits. In addition to the tertiary structure of the subunits, multiple-subunit proteins possess a quaternary structure, which is the arrangement into which the subunits assemble. Enzymes composed of subunits with diverse functions are sometimes called holoenzymes, in which some parts may be known as regulatory subunits and the functional core is known as the catalytic subunit. Examples of proteins with quaternary structure include hemoglobin, DNA polymerase, and ion channels. Other assemblies referred to instead as *multiprotein complexes* also possess quaternary structure. Examples include nucleosomes and microtubules. Changes in quaternary structure can occur through conformational changes within individual subunits or through reorientation of the subunits relative to each other. It is through such changes, which underlie cooperativity and allostery in "multimeric" enzymes, that many proteins undergo regulation and perform their physiological function.

The above definition follows a classical approach to biochemistry, established at times when the distinction between a protein and a functional, proteinaceous unit was difficult to elucidate. More recently, people refer to protein-protein interaction when discussing quaternary structure of proteins and consider all assemblies of proteins as protein complexes.

Nomenclature of Quaternary Structures

The number of subunits in an oligomeric complex are described using names that end in -mer. Formal and Greco-Latinate names are generally used for the first ten types and can be used for up to twenty subunits, whereas higher order complexes are usually described by the number of subunits, followed by -meric.

- 1 = monomer
- 2 = dimer
- 3 = trimer
- 4 = tetramer
- 5 = pentamer
- 6 = hexamer
- 7 = heptamer
- 8 = octamer
- 9 = nonamer
- 10 = decamer
- 11 = undecamer
- 12 = dodecamer
- 13 = tridecamer
- 14 = tetradecamer
- 15 = pentadecamer*
- 16 = hexadecamer
- 17 = heptadecamer*
- 18 = octadecamer
- 19 = nonadecamer
- 20 = eicosamer
- 21-mer
- 22-mer
- 23-mer* etc.

No known examples

Although complexes higher than octamers are rarely observed for most proteins, there are some important exceptions. Viral capsids are often composed of multiples of 60 proteins. Several molecular machines are also found in the cell, such as the proteasome (four heptameric rings = 28 subunits), the transcription complex and the spliceosome. The ribosome is probably the largest molecular machine, and is composed of many RNA and protein molecules.

In some cases, proteins form complexes that then assemble into even larger complexes. In such cases, one uses the nomenclature, e.g., "dimer of dimers" or "trimer of dimers", to suggest that the complex might dissociate into smaller sub-complexes before dissociating into monomers.

Determination of Quaternary Structure

Protein quaternary structure can be determined using a variety of experimental techniques that require a sample of protein in a variety of experimental conditions. The experiments often provide an estimate of the mass of the native protein and, together with knowledge of the masses and/or stoichiometry of the subunits, allow the quaternary structure to be predicted with a given accuracy. It is not always possible to obtain a precise determination of the subunit composition for a variety of reasons.

The number of subunits in a protein complex can often be determined by measuring the hydrodynamic molecular volume or mass of the intact complex, which requires native solution conditions. For *folded* proteins, the mass can be inferred from its volume using the partial specific volume of 0.73 ml/g. However, volume measurements are less certain than mass measurements, since *unfolded* proteins appear to have a much larger volume than folded proteins; additional experiments are required to determined whether a protein is unfolded or has formed an oligomer.

Prediction of Quaternary Structure Attribute

Some bioinformatics methods were developed for predicting the quaternary structural attributes of proteins based on their sequence information by using various modes of pseudo amino acid composition.

Methods that measure mass of intact complex directly:
- sedimentation-equilibrium analytical ultracentrifugation
- electrospray mass spectrometry
- Mass spectrometric immunoassay (MSIA)

Methods that measure the size of the intact complex directly:
- static light scattering
- size exclusion chromatography (requires calibration)
- Dual polarisation interferometry

Methods that measure the size of the intact complex indirectly:
- sedimentation-velocity analytical ultracentrifugation (measures the translational diffusion constant)
- dynamic light scattering (measures the translational diffusion constant)
- pulsed-gradient protein nuclear magnetic resonance (measures the translational diffusion constant)

- fluorescence polarization (measures the rotational diffusion constant)
- dielectric relaxation (measures the rotational diffusion constant)
- Dual polarisation interferometry (measures the size and the density of the complex)

Methods that measure the mass or volume under unfolding conditions (such as MALDI-TOF mass spectrometry and SDS-PAGE) are generally not useful, since non-native conditions usually cause the complex to dissociate into monomers. However, these may sometimes be applicable; for example, the experimenter may apply SDS-PAGE after first treating the intact complex with chemical cross-linking reagents.

Protein-protein Interactions

Proteins are capable of forming very tight complexes. For example, ribonuclease inhibitor binds to ribonuclease A with a roughly 20 fM dissociation constant. Other proteins have evolved to bind specifically to unusual moieties on another protein, e.g., biotin groups (avidin), phosphorylated tyrosines (SH2 domains) or proline-rich segments (SH3 domains).

Domains, Motifs, and Folds in Protein Structure

Protein are frequently described as consisting from several structural units.

A structural domain is an element of the protein's overall structure that is self-stabilizing and often folds independently of the rest of the protein chain. Many domains are not unique to the protein products of one gene or one gene family but instead appear in a variety of proteins. Domains often are named and singled out because they figure prominently in the biological function of the protein they belong to; for example, the "calcium-binding domain of calmodulin". Because they are independently stable, domains can be "swapped" by genetic engineering between one protein and another to make chimeras.

The structural and sequence motifs refer to short segments of protein three-dimensional structure or amino acid sequence that were found in a large number of different proteins.

The supersecondary structure refers to a specific combination of secondary structure elements, such as beta-alpha-beta units or helix-turn-helix motif. Some of them may be also referred to as structural motifs.

Protein fold refers to the general protein architecture, like helix bundle, beta-barrel, Rossman fold or different "folds" provided in the Structural Classification of Proteins database. Despite the fact that there are about 100,000 different proteins expressed in eukaryotic systems, there are many fewer different domains, structural motifs and folds. This is partly a consequence of evolution, since genes or parts of genes can be doubled or moved around within the genome.

This means that, for example, a protein domain might be moved from one protein to another thus giving the protein a new function. Because of these mechanisms, pathways and mechanisms tend to be reused in several different proteins.

Protein Folding

An unfolded polypeptide folds into its characteristic three-dimensional structure from random coil.

Protein Structure Determination

Around 90% of the protein structures available in the Protein Data Bank have been determined by X-ray crystallography. This method allows one to measure the 3D density distribution of electrons in the protein (in the crystallized state) and thereby infer the 3D coordinates of all the atoms to be determined to a certain resolution.

Roughly 9% of the known protein structures have been obtained by Nuclear Magnetic Resonance techniques. The secondary structure composition can be determined via circular dichroism or dual polarisation interferometry. Cryo-electron microscopy has recently become a means of determining protein structures to high resolution (less than 5 angstroms or 0.5 nanometer) and is anticipated to increase in power as a tool for high resolution work in the next decade. This technique is still a valuable resource for researchers working with very large protein complexes such as virus coat proteins and amyloid fibers.

Structure Classification

Protein structures can be classified based on their similarity or a common evolutionary origin. SCOP and CATH databases provide two different structural classifications of proteins.

Computational Prediction of Protein Structure

The generation of a protein sequence is much easier than the determination of a protein structure. However, the structure of a protein gives much more insight in the function of the protein than its sequence. Therefore, a number of methods for the computational prediction of protein structure from its sequence have been developed. *Ab initio* prediction methods use just the sequence of the protein. Threading and Homology Modeling methods can build a 3D model for a protein of unknown structure from experimental structures of evolutionary related proteins. Most proteins fold into unique 3-dimensional structures. The shape into which a protein naturally folds is known as its native conformation. Although many proteins can fold unassisted, simply through the chemical properties of their amino acids, others require the aid of molecular chaperones to fold into their native states. Biochemists often refer to four distinct aspects of a protein's structure:

- *Primary structure*: the amino acid sequence.
- *Secondary structure*: regularly repeating local structures stabilized by hydrogen bonds. The most common examples are the alpha helix, beta sheet and turns. Because secondary structures are local, many regions of different secondary structure can be present in the same protein molecule.
- *Tertiary structure*: the overall shape of a single protein molecule; the spatial relationship of the secondary structures to one another. Tertiary structure is generally stabilized by nonlocal interactions, most commonly the formation of a hydrophobic core, but also through salt bridges, hydrogen bonds, disulfide bonds, and even post-translational modifications. The term "tertiary structure" is often used as synonymous with the term *fold*. The tertiary structure is what controls the basic function of the protein.
- *Quaternary structure*: the structure formed by several protein molecules (polypeptide chains), usually called *protein subunits* in this context, which function as a single protein complex.

Proteins are not entirely rigid molecules. In addition to these levels of structure, proteins may shift between several related structures while they perform their functions. In the context of these functional rearrangements, these tertiary or quaternary structures are usually referred to as "conformations", and transitions between them are called *conformational changes*. Such changes are often induced by the binding of a substrate molecule to an enzyme's active site, or the physical region of the protein that participates in chemical catalysis.

In solution proteins also undergo variation in structure through thermal vibration and the collision with other molecules.

Proteins can be informally divided into three main classes, which correlate with typical tertiary structures: globular proteins, fibrous proteins, and membrane proteins. Almost all globular proteins are soluble and many are enzymes. Fibrous proteins are often structural, such as collagen, the major component of connective tissue, or keratin, the protein component of hair and nails. Membrane proteins often serve as receptors or provide channels for polar or charged molecules to pass through the cell membrane.

A special case of intramolecular hydrogen bonds within proteins, poorly shielded from water attack and hence promoting their own dehydration, are called dehydrons.

Structure Determination

Discovering the tertiary structure of a protein, or the quaternary structure of its complexes, can provide important clues about how the protein performs its function. Common experimental methods of structure determination include X-ray crystallography and NMR spectroscopy, both of which can produce information at atomic resolution. However, NMR experiments are able to provide information from which a subset of distances between pairs of atoms can be estimated, and the final possible conformations for a protein are determined by solving a distance geometry problem. Dual polarisation interferometry is a quantitative analytical method for measuring the overall protein conformation and conformational changes due to interactions or other stimulus. Circular dichroism is another laboratory technique for determining internal beta sheet/helical composition of proteins. Cryoelectron microscopy is used to produce lower-resolution structural information about very large protein complexes, including assembled viruses; a variant known as electron crystallography can also produce high-resolution information in some cases, especially for two-dimensional crystals of membrane proteins. Solved structures are usually deposited in the Protein Data Bank (PDB), a freely available resource from which structural data about thousands of proteins can be obtained in the form of Cartesian coordinates for each atom in the protein.

Many more gene sequences are known than protein structures. Further, the set of solved structures is biased toward proteins that can be easily subjected to the conditions required in X-ray crystallography, one of the major structure determination methods. In particular, globular proteins are comparatively easy to crystallize in preparation for X-ray crystallography. Membrane proteins, by contrast, are difficult to crystallize and are underrepresented in the PDB. Structural genomics initiatives have attempted to remedy these deficiencies by systematically solving representative structures of major fold classes. Protein structure prediction methods attempt to provide a means of generating a plausible structure for proteins whose structures have not been experimentally determined.

Cellular Functions

Proteins are the chief actors within the cell, said to be carrying out the duties specified by the information encoded in genes. With the exception of certain types of RNA, most other biological molecules are relatively inert elements upon which proteins act. Proteins make up half the dry weight of an *Escherichia coli* cell, whereas other macromolecules such as DNA and RNA make up only 3% and 20%, respectively. The set of proteins expressed in a particular cell or cell type is known as its proteome.

The chief characteristic of proteins that also allows their diverse set of functions is their ability to bind other molecules specifically and tightly. The region of the protein responsible for binding another molecule is known as the binding site and is often a depression or "pocket" on the molecular surface. This binding ability is mediated by the tertiary structure of the protein, which defines the binding site pocket, and by the chemical properties of the surrounding amino acids' side chains. Protein binding can be extraordinarily tight and specific; for example, the ribonuclease inhibitor protein binds to human angiogenin with a sub-femtomolar dissociation constant ($<10^{-5}$M) but does not bind at all to its amphibian homolog onconase (>1 M). Extremely minor chemical changes such as the addition of a single methyl group to a binding partner can sometimes suffice to nearly eliminate binding; for example, the aminoacyl tRNA synthetase specific to the amino acid valine discriminates against the very similar side chain of the amino acid isoleucine.

Proteins can bind to other proteins as well as to small-molecule substrates. When proteins bind specifically to other copies of the same molecule, they can oligomerize to form fibrils; this process occurs often in structural proteins that consist of globular monomers that self-associate to form rigid fibers. Protein–protein interactions also regulate enzymatic activity, control progression through the cell cycle, and allow the assembly of large protein complexes that carry out many closely related reactions with a common biological function. Proteins can also bind to, or even be integrated into, cell membranes. The ability of binding partners to induce conformational changes in proteins allows the construction of enormously complex signalling networks. Importantly, as interactions between proteins are reversible, and depend heavily on the availability of different groups of partner proteins to form aggregates that are capable to carry out discrete sets of function, study of the interactions between specific proteins is a key to understand important aspects of cellular function, and ultimately the properties that distinguish particular cell types.

Enzymes

The best-known role of proteins in the cell is as enzymes, which catalyse chemical reactions. Enzymes are usually highly specific and accelerate only one or a few chemical reactions. Enzymes carry out most of the reactions involved in metabolism, as well as manipulating DNA in processes such as DNA replication, DNA repair, and transcription. Some enzymes act on other proteins to add or remove chemical groups in a process known as post-translational modification. About 4,000 reactions are known to be catalysed by enzymes. The rate acceleration conferred by enzymatic catalysis is often enormous—as much as 10^5-fold increase in rate over the uncatalyzed reaction in the case of orotate decarboxylase (78 million years without the enzyme, 18 milliseconds with the enzyme).

The molecules bound and acted upon by enzymes are called substrates. Although enzymes can consist of hundreds of amino acids, it is usually only a small fraction of the residues that come in contact with the substrate, and an even smaller fraction—three to four residues on average—that are directly involved in catalysis. The region of the enzyme that binds the substrate and contains the catalytic residues is known as the active site.

Cell Signalling and Ligand Binding

Many proteins are involved in the process of cell signalling and signal transduction. Some proteins, such as insulin, are extracellular proteins that transmit a signal from the cell in which they were synthesized to other cells in distant tissues. Others are membrane proteins that act as receptors whose main function is to bind a signalling molecule and induce a biochemical response in the cell. Many receptors have a binding site exposed on the cell surface and an effector domain within the cell, which may have enzymatic activity or may undergo a conformational change detected by other proteins within the cell.

Antibodies are protein components of adaptive immune system whose main function is to bind antigens, or foreign substances in the body, and target them for destruction. Antibodies can be secreted into the extracellular environment or anchored in the membranes of specialized B cells known as plasma cells. Whereas enzymes are limited in their binding affinity for their substrates by the necessity of conducting their reaction, antibodies have no such constraints. An antibody's binding affinity to its target is extraordinarily high.

Many ligand transport proteins bind particular small biomolecules and transport them to other locations in the body of a multicellular organism. These proteins must have a high binding affinity when their ligand is present in high concentrations, but must also release the ligand when it is present at low concentrations in the target tissues. The canonical example of a ligand-binding protein is haemoglobin, which transports oxygen from the lungs to other organs and tissues in all vertebrates and has close homologs in every biological kingdom. Lectins are sugar-binding proteins which are highly specific for their sugar moieties. Lectins typically play a role in biological recognition phenomena involving cells and proteins. Receptors and hormones are highly specific binding proteins.

Transmembrane proteins can also serve as ligand transport proteins that alter the permeability of the cell membrane to small molecules and ions. The membrane alone has a hydrophobic core through which polar or charged molecules cannot diffuse. Membrane proteins contain internal channels that allow such molecules to enter and exit the cell. Many ion channel proteins are specialized to select for only a particular ion; for example, potassium and sodium channels often discriminate for only one of the two ions.

Structural Proteins

Structural proteins confer stiffness and rigidity to otherwise-fluid biological components. Most structural proteins are fibrous proteins; for example, actin and tubulin are globular and soluble as monomers, but polymerize to form long, stiff fibers that comprise the cytoskeleton, which allows the cell to maintain its shape and size. Collagen and elastin are critical components of connective tissue such as cartilage, and keratin is found in hard or filamentous structures such as hair, nails, feathers, hooves, and some animal shells. Other proteins that serve structural functions are motor proteins such as myosin, kinesin, and dynein, which are capable of generating mechanical forces. These proteins are crucial for cellular motility of single celled organisms and the sperm of many multicellular organisms which reproduce sexually. They also generate the forces exerted by contracting muscles.

Cellular Localization

The study of proteins *in vivo* is often concerned with the synthesis and localization of the protein within the cell. Although many intracellular proteins are synthesized in the cytoplasm and membrane-bound or secreted proteins in the endoplasmic reticulum, the specifics of how proteins are targeted to specific organelles or cellular structures is often unclear. A useful technique for assessing cellular localization uses genetic engineering to express in a cell a fusion protein or chimera consisting of the natural protein of interest linked to a "reporter" such as green fluorescent protein (GFP). The fused protein's position within the cell can be cleanly and efficiently visualized using microscopy, as shown in the figure opposite.

Other methods for elucidating the cellular location of proteins requires the use of known compartmental markers for regions such as the ER, the Golgi, lysosomes/vacuoles, mitochondria, chloroplasts, plasma membrane, etc. With the use of fluorescently tagged versions of these markers or of antibodies to known markers, it becomes much simpler to identify the localization of a protein of interest. For example, indirect immunofluorescence will allow for fluorescence colocalization and demonstration of location. Fluorescent dyes are used to label cellular compartments for a similar purpose.

Other possibilities exist, as well. For example, immunohistochemistry usually utilizes an antibody to one or more proteins of interest that are conjugated to enzymes yielding either luminescent or chromogenic signals that can be compared between samples, allowing for localization information. Another applicable technique is cofractionation in sucrose (or other material) gradients using isopycnic centrifugation. While this technique does not prove colocalization of a compartment of known density and the protein of interest, it does increase the likelihood, and is more amenable to large-scale studies.

Finally, the gold-standard method of cellular localization is immunoelectron microscopy. This technique also uses an antibody to the protein of interest, along with classical electron microscopy techniques. The sample is prepared for normal electron microscopic examination, and then treated with an antibody to the protein of interest that is conjugated to an extremely electro-dense material, usually gold. This allows for the localization of both ultrastructural details as well as the protein of interest. Through another genetic engineering application known as site-directed mutagenesis, researchers can alter the protein sequence and hence its structure, cellular localization, and susceptibility to regulation. This technique even allows the incorporation of unnatural amino acids into proteins, using modified tRNAs, and may allow the rational design of new proteins with novel properties.

Proteomics and Bioinformatics

The total complement of proteins present at a time in a cell or cell type is known as its proteome, and the study of such large-scale data sets defines the field of proteomics, named by analogy to the related field of genomics. Key experimental techniques in proteomics include 2D electrophoresis, which allows the separation of a large number of proteins, mass spectrometry, which allows rapid high-throughput identification of proteins and sequencing of peptides (most often after in-gel digestion), protein microarrays, which allow the detection of the relative levels of a large number of proteins present in a cell, and two-hybrid screening, which allows the systematic exploration of protein–protein interactions.

The total complement of biologically possible such interactions is known as the interactome. A systematic attempt to determine the structures of proteins representing every possible fold is known as structural genomics.

The large amount of genomic and proteomic data available for a variety of organisms, including the human genome, allows researchers to efficiently identify homologous proteins in distantly related organisms by sequence alignment. Sequence profiling tools can perform more specific sequence manipulations such as restriction enzyme maps, open reading frame analyses for nucleotide sequences, and secondary structure prediction. From this data phylogenetic trees can be constructed and evolutionary hypotheses developed using special software like ClustalW regarding the ancestry of modern organisms and the genes they express. The field of bioinformatics seeks to assemble, annotate, and analyze genomic and proteomic data, applying computational techniques to biological problems such as gene finding and cladistics.

Structure Prediction and Simulation

Complementary to the field of structural genomics, protein structure prediction seeks to develop efficient ways to provide plausible models for proteins whose structures have not yet been determined experimentally.

The most successful type of structure prediction, known as homology modeling, relies on the existence of a "template" structure with sequence similarity to the protein being modeled; structural genomics' goal is to provide sufficient representation in solved structures to model most of those that remain.

Although producing accurate models remains a challenge when only distantly related template structures are available, it has been suggested that sequence alignment is the bottleneck in this process, as quite accurate models can be produced if a "perfect" sequence alignment is known.

Many structure prediction methods have served to inform the emerging field of protein engineering, in which novel protein folds have already been designed. A more complex computational problem is the prediction of intermolecular interactions, such as in molecular docking and protein–protein interaction prediction.

The processes of protein folding and binding can be simulated using such technique as molecular mechanics, in particular, molecular dynamics and Monte Carlo, which increasingly take advantage of parallel and distributed computing (Folding@Home project; molecular modeling on GPU).

The folding of small alpha-helical protein domains such as the villin headpiece and the HIV accessory protein have been successfully simulated *in silico*, and hybrid methods that combine standard molecular dynamics with quantum mechanics calculations have allowed exploration of the electronic states of rhodopsins.

Nutrition

Most microorganisms and plants can biosynthesize all 20 standard amino acids, while animals (including humans) must obtain some of the amino acids from the diet. The amino acids that an organism cannot synthesize on its own are referred to as essential amino acids. Key enzymes that synthesize certain amino acids are not present in animals — such as aspartokinase, which catalyzes the first step in the synthesis of lysine, methionine, and threonine from aspartate.

If amino acids are present in the environment, microorganisms can conserve energy by taking up the amino acids from their surroundings and downregulating their biosynthetic pathways. In animals, amino acids are obtained through the consumption of foods containing protein. Ingested proteins are then broken down into amino acids through digestion, which typically involves denaturation of the protein through exposure to acid and hydrolysis by enzymes called proteases. Some ingested amino acids are used for protein biosynthesis, while others are converted to glucose through gluconeogenesis, or fed into the citric acid cycle.

This use of protein as a fuel is particularly important under starvation conditions as it allows the body's own proteins to be used to support life, particularly those found in muscle. Amino acids are also an important dietary source of nitrogen.

Peptide Bond

A peptide bond (amide bond) is a covalent chemical bond formed between two molecules when the carboxyl group of one molecule reacts with the amino group of the other molecule, thereby releasing a molecule of water (H_2O).

This is a dehydration synthesis reaction (also known as a condensation reaction), and usually occurs between amino acids. The resulting C(O)NH bond is called a peptide bond, and the resulting molecule is an amide. The four-atom functional group -C(=O)NH- is called a peptide link. Polypeptides and proteins are chains of amino acids held together by peptide bonds, as is the backbone of PNA.

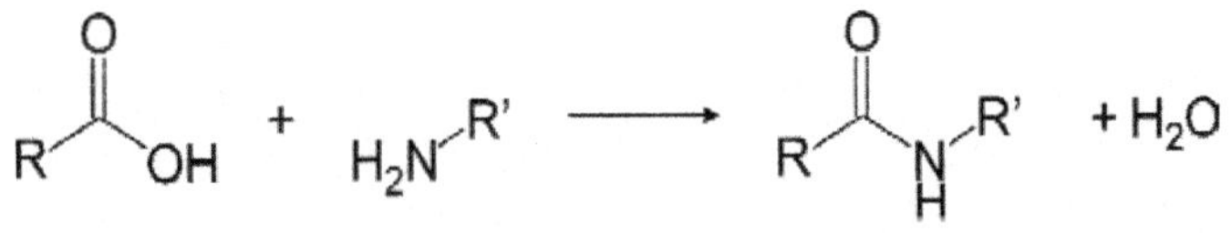

Figure 1: Dehydration synthesis (condensation) reaction forming an amide

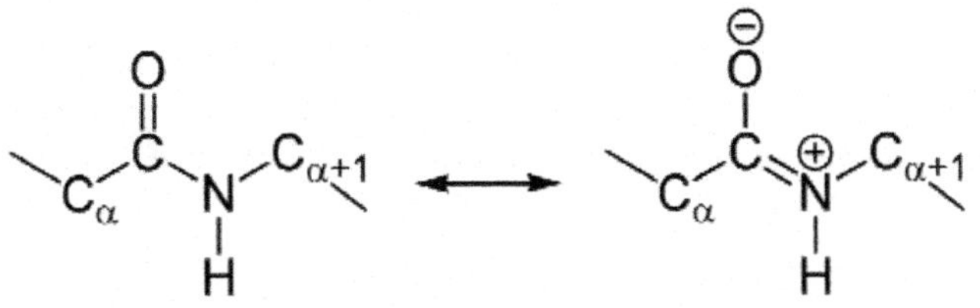

Figure 2: Resonance forms of a typical peptide group. The uncharged, single-bonded form (typically ~60%) is shown on the left, whereas the charged, double-bonded form (typically ~40%) is on the right.

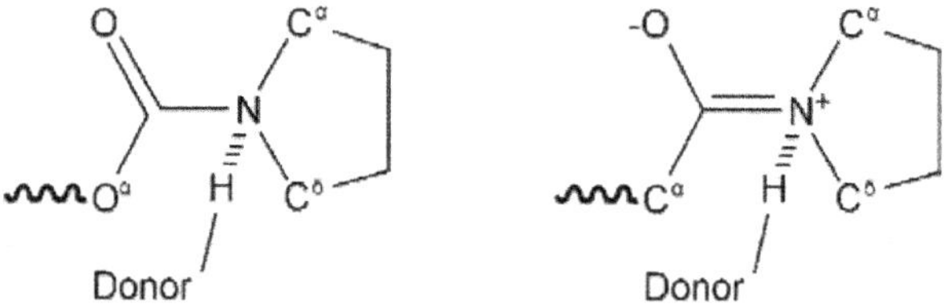

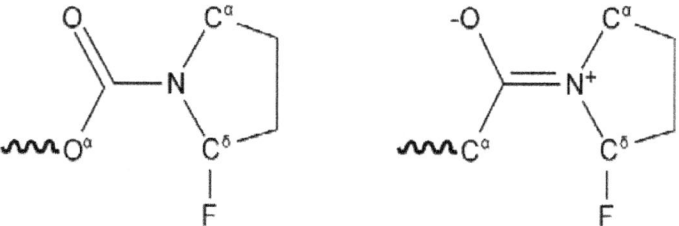

Figure 3: A hydrogen bond (dotted line) to an X-Pro peptide group favors the single-bonded resonance form (left) over the double-bonded form (right).

Figure 4: An electronegative substituent near the amide nitrogen favors the single-bonded resonance form (left) over the double-bonded form (right).

A peptide bond can be broken by amide hydrolysis (the adding of water). The peptide bonds in proteins are metastable, meaning that in the presence of water they will break spontaneously, releasing 2-4 kcal/mol of free energy, but this process is extremely slow.

In living organisms, the process is facilitated by enzymes. Living organisms also employ enzymes to form peptide bonds; this process requires free energy. The wavelength of absorbance for a peptide bond is 190-230 nm.

Resonance Forms of the Peptide Group

The amide group has two resonance forms, which confer several important properties. First, it stabilizes the group by roughly 20 kcal/mol, making it less reactive than many similar groups (such as esters). The resonance suggests that the amide group has a partial double bond character, estimated at 40% under typical conditions. The peptide group is uncharged at all normal pH values, but its double-bonded resonance form gives it an unusually large dipole moment, roughly 3.5 Debye (0.7 electron-angstrom). These dipole moments can line up in certain secondary structures (such as the α-helix), producing a large net dipole.

The partial double bond character can be strengthened or weakened by modifications that favour one resonance form over another. For example, the double-bonded form is disfavored in hydrophobic environments, because of its charge. Conversely, donating a hydrogen bond to the amide oxygen or accepting a hydrogen bond from the amide nitrogen should favour the double-bonded form, because the hydrogen bond should be stronger to the charged form than to the uncharged, single-bonded form.

By contrast, donating a hydrogen bond to an amide nitrogen in an X-Pro peptide bond should favour the single-bonded form; donating it to the double-bonded form would give the nitrogen five quasi-covalent bonds! Similarly, a strongly electronegative substituent (such as fluorine) near the amide nitrogen favors the single-bonded form, by competing with the amide oxygen to "steal" an electron from the amide nitrogen.

Cis/trans Isomers of the Peptide Group

The partial double bond renders the amide group planar, occurring in either the cis or trans isomers. In the unfolded state of proteins, the peptide groups are free to isomerize and adopt both isomers; however, in the folded state, only a single isomer is adopted at each position (with rare exceptions). The trans form is preferred overwhelmingly in most peptide bonds (roughly 1000:1 ratio in trans:cis populations). However, X-Pro peptide groups tend to have a roughly 3:1 ratio, presumably because the symmetry between the C^α and C^δ atoms of proline makes the cis and trans isomers nearly equal in energy.

The dihedral angle associated with the peptide group (defined by the four atoms $c^\alpha - C' - N - C^\alpha$) is denoted ω; $\omega = 0°$ for the cis isomer and $\omega = 180°$ for the trans isomer. Amide groups can isomerize about the C-N bond between the cis and trans forms, albeit slowly $\tau \sim$ (20 seconds at room temperature). The transition states $\omega = \pm 90°$ requires that the partial double bond be broken, so that the activation energy is roughly 20 kcal/mol. However, the activation energy can be lowered (and the isomerization catalysed) by changes that favour the single-bonded form, such as placing the peptide group in a hydrophobic environment or donating a hydrogen bond to the nitrogen atom of an X-Pro peptide group. Both of these mechanisms for lowering the activation energy have been observed in peptidyl prolyl isomerases (PPIases), which are naturally occurring enzymes that catalyse the cis-trans isomerization of X-Pro peptide bonds.

Conformational protein folding is usually much faster (typically 10-100 ms) than cis-trans isomerization (10-100 s). A nonnative isomer of some peptide groups can disrupt the conformational folding significantly, either slowing it or preventing it from even occurring until the native isomer is reached. However, not all peptide groups have the same effect on folding; nonnative isomers of other peptide groups may not affect folding at all.

Chemical Reactions

Due to its resonance stabilization, the peptide bond is relatively unreactive under physiological conditions, even less than similar compounds such as esters. Nevertheless, peptide bonds can undergo chemical reactions, usually through an attack of an electronegative atom on the carbonyl carbon, breaking the carbonyl double bond and forming a tetrahedral intermediate. This is the pathway followed in proteolysis and, more generally, in N-O acyl exchange reactions such as those of inteins. When the functional group attacking the peptide bond is a thiol, hydroxyl or amine, the resulting molecule may be called a cyclol or, more specifically, a thiacyclol, an oxacyclol or an azacyclol, respectively.

4

Biological and Chemical Organization of Cell

The cell is the functional basic unit of life. It was discovered by Robert Hooke and is the functional unit of all known living organisms. It is the smallest unit of life that is classified as a living thing, and is often called the building block of life. Some organisms, such as most bacteria, are unicellular (consist of a single cell). Other organisms, such as humans, are multicellular. Humans have about 100 trillion or 10^{14} cells; a typical cell size is 10 µm and a typical cell mass is 1 nanogram. The largest cells are about 135 µm in the anterior horn in the spinal cord while granule cells in the cerebellum, the smallest, can be some 4 µm and the longest cell can reach from the toe to the lower brain stem (Pseudounipolar cells). The largest known cells are unfertilised ostrich egg cells which weigh 3.3 pounds.

In 1835, before the final cell theory was developed, Jan Evangelista Purkynì observed small "granules" while looking at the plant tissue through a microscope. The cell theory, first developed in 1839 by Matthias Jakob Schleiden and Theodor Schwann, states that all organisms are composed of one or more cells, that all cells come from preexisting cells, that vital functions of an organism occur within cells, and that all cells contain the hereditary information necessary for regulating cell functions and for transmitting information to the next generation of cells.

The word *cell* comes from the Latin *cellula*, meaning, a small room. The descriptive term for the smallest living biological structure was coined by Robert Hooke in a book he published in 1665 when he compared the cork cells he saw through his microscope to the small rooms monks lived in.

Types of Cell

Eukaryotic and Prokaryotic, two types of cell.

Prokaryotic Cells

Prokaryotic cells are usually independent, this cell is simpler, and therefore smaller, than a eukaryote cell, lacking a nucleus and most of the other organelles of eukaryotes. There are two kinds of prokaryotes: bacteria and archaea; these share a similar structure.

Nuclear material of prokaryotic cell consist of a single chromosome which is in direct contact with cytoplasm. Here the undefined nuclear region in the cytoplasm is called nucleoid.

A prokaryotic cell has three architectural regions:

- On the outside, flagella and pili project from the cell's surface. These are structures (not present in all prokaryotes) made of proteins that facilitate movement and communication between cells;

- Enclosing the cell is the cell envelope – generally consisting of a cell wall covering a plasma membrane though some bacteria also have a further covering layer called a capsule. The envelope gives rigidity to the cell and separates the interior of the cell from its environment, serving as a protective filter. Though most prokaryotes have a cell wall, there are exceptions such as *Mycoplasma* (bacteria) and *Thermoplasma* (archaea). The cell wall consists of *peptidoglycan* in bacteria, and acts as an additional barrier against exterior forces. It also prevents the cell from expanding and finally bursting (cytolysis) from osmotic pressure against a hypotonic environment. Some eukaryote cells (plant cells and fungi cells) also have a cell wall;

- Inside the cell is the cytoplasmic region that contains the cell genome (DNA) and ribosomes and various sorts of inclusions. A prokaryotic chromosome is usually a circular molecule (an exception is that of the bacterium *Borrelia burgdorferi*, which causes Lyme disease). Though not forming a *nucleus*, the DNA is condensed in a *nucleoid*. Prokaryotes can carry extrachromosomal DNA elements called *plasmids*, which are usually circular. Plasmids enable additional functions, such as antibiotic resistance.

Eukaryotic Cells

Eukaryotic cells are often found in multicellular organisms. These cells are about 15 times wider than a typical prokaryote and can be as much as 1000 times greater in volume. The major difference between prokaryotes and eukaryotes is that eukaryotic cells contain membrane-bound compartments in which specific metabolic activities take place.

Most important among these is a cell nucleus, a membrane-delineated compartment that houses the eukaryotic cell's DNA. This nucleus gives the eukaryote its name, which means "true nucleus." Other differences include:

- The plasma membrane resembles that of prokaryotes in function, with minor differences in the setup. Cell walls may or may not be present.

- The eukaryotic DNA is organized in one or more linear molecules, called chromosomes, which are associated with histone proteins. All chromosomal DNA is stored in the *cell nucleus*, separated from the cytoplasm by a membrane. Some eukaryotic organelles such as mitochondria also contain some DNA.

- Many eukaryotic cells are ciliated with *primary cilia*. Primary cilia play important roles in chemosensation, mechanosensation, and thermosensation. Cilia may thus be "viewed as sensory cellular antennae that coordinate a large number of cellular signalling pathways, sometimes coupling the signalling to ciliary motility or alternatively to cell division and differentiation."

- Eukaryotes can move using *motile cilia* or *flagella*. The flagella are more complex than those of prokaryotes.

Subcellular Components

All cells, whether prokaryotic or eukaryotic, have a membrane that envelops the cell, separates its interior from its environment, regulates what moves in and out (selectively permeable), and maintains the electric potential of the cell. Inside the membrane, a salty cytoplasm takes up most of the cell volume. All cells possess DNA, the hereditary material of genes, and RNA, containing the information necessary to build various proteins such as enzymes, the cell's primary machinery. There are also other kinds of biomolecules in cells. This article will list these primary components of the cell, then briefly describe their function.

Membrane

The cell membrane is a biological membrane that separates the interior of all cells from the outside environment.. The cell membrane is selectively-permeable to ions and organic molecules and controls the movement of substances in and out of cells.. It consists of the phospholipid bilayer, consisting of hydrophobic tails and hydrophilic heads with embedded proteins, which are involved in a variety of cellular processes such as cell adhesion, ion conductivity and cell signalling. The plasma membrane also serves as the attachment surface for the extracellular glycocalyx and cell wall and intracellular cytoskeleton.

Function

The cell membrane surrounds the protoplasm of a cell and, in animal cells, physically separates the intracellular components from the extracellular environment. Fungi, bacteria and plants also have the cell wall which provides a mechanical support for the cell and precludes passage of the larger molecules. The cell membrane also plays a role in anchoring the cytoskeleton to provide shape to the cell, and in attaching to the extracellular matrix and other cells to help group cells together to form tissues. The barrier is differentially permeable and able to regulate what enters and exits the cell, thus facilitating the transport of materials needed for survival. The movement of substances across the membrane can be either *passive*, occurring without the input of cellular energy, or active, requiring the cell to expend energy in moving it. The membrane also maintains the cell potential.

Structure

Fluid Mosaic Model: According to the fluid mosaic model of S. J. Singer and Garth Nicolson 1972, the biological membranes can be considered as a two-dimensional liquid where all lipid and protein molecules diffuse more or less easily. This picture may be valid in the space scale of 10 nm. However, the plasma membranes contain different structures or domains that can be classified as: (a) protein-protein complexes; (b) lipid rafts, and (c) pickets and fences formed by the actin-based cytoskeleton.

Lipid Bilayer

The cell membrane consists primarily of a thin layer of amphipathic phospholipids which spontaneously arrange so that the hydrophobic "tail" regions are shielded from the surrounding polar fluid, causing the more hydrophilic "head" regions to associate with the cytosolic and extracellular faces of the resulting bilayer. This forms a continuous, spherical lipid bilayer.

The arrangement of hydrophilic heads and hydrophobic tails of the lipid bilayer prevent polar solutes (e.g. amino acids, nucleic acids, carbohydrates, proteins, and ions) from diffusing across the membrane, but generally allows for the passive diffusion of hydrophobic molecules. This affords the cell the ability to control the movement of these substances via transmembrane protein complexes such as pores and gates. Flippases and Scramblases concentrate phosphatidyl serine, which carries a negative charge, on the inner membrane. Along with NANA, this creates an extra barrier to charged moieties moving through the membrane.

Membranes serve diverse functions in eukaryotic and prokaryotic cells. One important role is to regulate the movement of materials into and out of cells. The phospholipid bilayer structure (fluid mosaic model) with specific membrane proteins accounts for the selective permeability of the membrane and passive and active transport mechanisms. In addition, membranes in prokaryotes and in the mitochondria and chloroplasts of eukaryotes facilitate the synthesis of ATP through chemiosmosis.

Membrane Polarity

The apical membrane of a polarized cell is the surface of the plasma membrane that faces the lumen. This is particularly evident in epithelial and endothelial cells, but also describes other polarized cells, such as neurons.

The basolateral membrane of a polarized cell is the surface of the plasma membrane that forms its basal and lateral surfaces. It faces towards the interstitium, and away from the lumen.

"Basolateral membrane" is a compound phrase referring to the terms *basal (base) membrane* and *lateral (side) membrane*, which, especially in epithelial cells, are identical in composition and activity. Proteins (such as ion channels and pumps) are free to move from the basal to the lateral surface of the cell or *vice versa* in accordance with the fluid mosaic model.

Tight junctions that join epithelial cells near their apical surface prevent the migration of proteins from the basolateral membrane to the apical membrane. The basal and lateral surfaces thus remain roughly equivalent to one another, yet distinct from the apical surface.

Integral Membrane Proteins

The cell membrane contains many integral membrane proteins, which pepper the entire surface. These structures, which can be visualized by electron microscopy or fluorescence microscopy, can be found on the inside of the membrane, the outside, or membrane spanning. These may include integrins, cadherins, desmosomes, clathrin-coated pits, caveolaes, and different structures involved in cell adhesion.

Membrane Skeleton

The cytoskeleton is found underlying the cell membrane in the cytoplasm and provides a scaffolding for membrane proteins to anchor to, as well as forming organelles that extend from the cell. Indeed, cytoskeletal elements interact extensively and intimately with the cell membrane. Anchoring proteins restricts them to a particular cell surface — for example, the *apical surface* of epithelial cells that line the vertebrate gut — and limits how far they may diffuse within the bilayer. The cytoskeleton is able to form appendage-like organelles, such as cilia, which are microtubule-based extensions covered by the cell membrane, and filopodia, which are actin-based extensions. These extensions are ensheathed in membrane and project from the surface of the cell in order to sense the external environment and/or make contact with the substrate or other cells.

The apical surfaces of epithelial cells are dense with actin-based finger-like projections known as microvilli, which increase cell surface area and thereby increase the absorption rate of nutrients. Localized decoupling of the cytoskeleton and cell membrane results in formation of a bleb.

Composition

Cell membranes contain a variety of biological molecules, notably lipids and proteins. Material is incorporated into the membrane, or deleted from it, by a variety of mechanisms:

- Fusion of intracellular vesicles with the membrane (exocytosis) not only excretes the contents of the vesicle but also incorporates the vesicle membrane's components into the cell membrane. The membrane may form blebs around extracellular material that pinch off to become vesicles (endocytosis).

- If a membrane is continuous with a tubular structure made of membrane material, then material from the tube can be drawn into the membrane continuously.

- Although the concentration of membrane components in the aqueous phase is low (stable membrane components have low solubility in water), there is an exchange of molecules between the lipid and aqueous phases.

Lipids

The cell membrane consists of three classes of amphipathic lipids: phospholipids, glycolipids, and cholesterols. The amount of each depends upon the type of cell, but in the majority of cases phospholipids are the most abundant. In RBC studies, 30% of the plasma membrane is lipid.

The fatty chains in phospholipids and glycolipids usually contain an even number of carbon atoms, typically between 16 and 20. The 16- and 18-carbon fatty acids are the most common. Fatty acids may be saturated or unsaturated, with the configuration of the double bonds nearly always *cis*.

The length and the degree of unsaturation of fatty acid chains have a profound effect on membrane fluidity as unsaturated lipids create a kink, preventing the fatty acids from packing together as tightly, thus decreasing the melting temperature (increasing the fluidity) of the membrane. The ability of some organisms to regulate the fluidity of their cell membranes by altering lipid composition is called homeoviscous adaptation.

The entire membrane is held together via non-covalent interaction of hydrophobic tails, however the structure is quite fluid and not fixed rigidly in place. Under physiological conditions phospholipid molecules in the cell membrane are in the liquid crystalline state. It means the lipid molecules are free to diffuse and exhibit rapid lateral diffusion along the layer in which they are present. However, the exchange of phospholipid molecules between intracellular and extracellular leaflets of the bilayer is a very slow process. Lipid rafts and caveolae are examples of cholesterol-enriched microdomains in the cell membrane.

In animal cells cholesterol is normally found dispersed in varying degrees throughout cell membranes, in the irregular spaces between the hydrophobic tails of the membrane lipids, where it confers a stiffening and strengthening effect on the membrane.

Carbohydrates

Plasma membranes also contain carbohydrates, predominantly glycoproteins, but with some glycolipids (cerebrosides and gangliosides). For the most part, no glycosylation occurs on membranes within the cell; rather generally glycosylation occurs on the extracellular surface of the plasma membrane.

The glycocalyx is an important feature in all cells, especially epithelia with microvilli. Recent data suggest the glycocalyx participates in cell adhesion, lymphocyte homing, and many others.

The penultimate sugar is galactose and the terminal sugar is sialic acid, as the sugar backbone is modified in the golgi apparatus. Sialic acid carries a negative charge, providing an external barrier to charged particles.

Proteins

Type	Description	Examples
Integral proteins or *transmembrane proteins*	Span the membrane and have a hydrophilic cytosolic domain, which interacts with internal molecules, a hydrophobic membrane spanning domain that anchors it within the cell membrane, and a hydrophilic extracellular domain that interacts with external molecules. The hydrophobic domain consists of one, multiple, or a combination of α-helices and β sheet protein motifs.	Ion channels, proton pumps, G protein-coupled receptor
Lipid anchored proteins	Covalently-bound to single or multiple lipid molecules; hydrophobically insert into the cell membrane and anchor the protein. The protein itself is not in contact with the membrane.	G proteins
Peripheral proteins	Attached to integral membrane proteins, or associated with peripheral regions of the lipid bilayer. These proteins tend to have only temporary interactions with biological membranes, and, once reacted the molecule, dissociates to carry on its work in the cytoplasm.	Some enzymes, some hormones

The cell membrane plays host to a large amount of protein that is responsible for its various activities. The amount of protein differs between species and according to function, however the typical amount in a cell membrane is 50%. These proteins are undoubtedly important to a cell: Approximately a third of the genes in yeast code specifically for them, and this number is even higher in multicellular organisms.

The cell membrane, being exposed to the outside environment, is an important site of cell-cell communication. As such, a large variety of protein receptors and identification proteins, such as antigens, are present on the surface of the membrane. Functions of membrane proteins can also include cell-cell contact, surface recognition, cytoskeleton contact, signalling, enzymatic activity, or transporting substances across the membrane.

Most membrane proteins must be inserted in some way into the membrane. For this to occur, an N-terminus "signal sequence" of amino acids directs proteins to the endoplasmic reticulum, which inserts the proteins into a lipid bilayer. Once inserted, the proteins are then transported to their final destination in vesicles, where the vesicle fuses with the target membrane.

Variation

The cell membrane has different lipid and protein compositions in distinct types of cells and may have therefore specific names for certain cell types:

- Sarcolemma in myocytes
- Oolemma in oocytes.

Permeability

The permeability of a membrane is the ease of molecules to pass through it. Permeability depends mainly on the electric charge of the molecule and to a lesser extent the molar mass of the molecule. Electrically neutral and small molecules pass the membrane easier than charged, large ones. The inability of charged molecules to pass through the cell membrane results in pH parturition of substances throughout the fluid compartments of the body.

Cytoskeleton

The cytoskeleton (also CSK) is a cellular "scaffolding" or "skeleton" contained within the cytoplasm and is made out of protein. The cytoskeleton is present in all cells; it was once thought to be unique to eukaryotes, but recent research has identified the prokaryotic cytoskeleton. It has structures such as flagella, cilia and lamellipodia and plays important roles in both intracellular transport (the movement of vesicles and organelles, for example) and cellular division. The concept of a protein mosaic that dynamically coordinated cytoplasmic biochemistry was proposed by Rudolph Peters in 1929 while the term (*cytosquelette*, in French) was first introduced by French embryologist Paul Wintrebert in 1931. It is almost like the human skeleton.

The Eukaryotic Cytoskeleton

Eukaryotic cells contain three main kinds of cytoskeletal filaments, which are microfilaments, intermediate filaments, and microtubules. The cytoskeleton provides the cell with structure and shape, and by excluding macromolecules from some of the cytosol it adds to the level of macromolecular crowding in this compartment. Cytoskeletal elements interact extensively and intimately with cellular membranes.

Intermediate Filaments

These filaments, around 10 manometers in diameter, are more stable (strongly bound) than actin filaments, and heterogeneous constituents of the cytoskeleton. Although little work has been done on intermediate filaments in plants, there is some evidence that cytosolic intermediate filaments might be present, and plant nuclear filaments have been detected. Like actin filaments, they function in the maintenance of cell-shape by bearing tension (microtubules, by contrast, resist compression. It may be useful to think of micro- and intermediate filaments as cables, and of microtubules as cellular support beams). Intermediate filaments organize the internal tridimensional structure of the cell, anchoring organelles and serving as structural components of the nuclear lamina and sarcomeres. They also participate in some cell-cell and cell-matrix junctions.

Different intermediate filaments are:

- made of vimentins, being the common structural support of many cells.
- made of keratin, found in skin cells, hair and nails.
- neurofilaments of neural cells.
- made of lamin, giving structural support to the nuclear envelope.

Microtubules

Microtubules are hollow cylinders about 23 nm in diameter (lumen = approximately 15nm in diameter), most commonly comprising 13 protofilaments which, in turn, are polymers of alpha and beta tubulin. They have a very dynamic behaviour, binding GTP for polymerization. They are commonly organized by the centrosome.

In nine triplet sets (star-shaped), they form the centrioles, and in nine doublets oriented about two additional microtubules (wheel-shaped) they form cilia and flagella. The latter formation is commonly referred to as a "9+2" arrangement, wherein each doublet is connected to another by the protein dynein. As both flagella and cilia are structural components of the cell, and are maintained by microtubules, they can be considered part of the cytoskeleton.

They play key roles in:

- intracellular transport (associated with dyneins and kinesins, they transport organelles like mitochondria or vesicles).
- the axoneme of cilia and flagella.
- the mitotic spindle.
- synthesis of the cell wall in plants.

The Prokaryotic Cytoskeleton

The cytoskeleton was previously thought to be a feature only of eukaryotic cells, but homologues to all the major proteins of the eukaryotic cytoskeleton have recently been found in prokaryotes. Although the evolutionary relationships are so distant that they are not obvious from protein sequence comparisons alone, the similarity of their three-dimensional structures and similar functions in maintaining cell shape and polarity provides strong evidence that the eukaryotic and prokaryotic cytoskeletons are truly homologous. However, some structures in the bacterial cytoskeleton may have yet to be identified.

FtsZ

FtsZ was the first protein of the prokaryotic cytoskeleton to be identified. Like tubulin, FtsZ forms filaments in the presence of GTP, but these filaments do not group into tubules. During cell division, FtsZ is the first protein to move to the division site, and is essential for recruiting other proteins that synthesize the new cell wall between the dividing cells.

MreB and ParM

Prokaryotic actin-like proteins, such as MreB, are involved in the maintenance of cell shape. All non-spherical bacteria have genes encoding actin-like proteins, and these proteins form a helical network beneath the cell membrane that guides the proteins involved in cell wall biosynthesis.

Some plasmids encode a partitioning system that involves an actin-like protein ParM. Filaments of ParM exhibit dynamic instability, and may partition plasmid DNA into the dividing daughter cells by a mechanism analogous to that used by microtubules during eukaryotic mitosis.

Crescentin

The bacterium *Caulobacter crescentus* contains a third 3rd protein, crescentin, that is related to the intermediate filaments of eukaryotic cells. Crescentin is also involved in maintaining cell shape, such as helical and vibrioid forms of bacteria, but the mechanism by which it does this is currently unclear.

History

Microtrabeculae

A fourth eukaryotic cytoskeletal element, *microtrabeculae*, was proposed by Keith Porter based on images obtained from high-voltage electron microscopy of whole cells in the 1970s. Porter proposed that this microtrabecular structure represented a novel filamentous network distinct from microtubules, filamentous actin, or intermediate filaments. It is now generally accepted that microtrabeculae are nothing more than an artifact of certain types of fixation treatment, although we have yet to fully understand the complexity of the cell's cytoskeleton.

The cytoskeleton acts to organize and maintain the cell's shape; anchors organelles in place; helps during endocytosis, the uptake of external materials by a cell, and cytokinesis, the separation of daughter cells after cell division; and moves parts of the cell in processes of growth and mobility. The eukaryotic cytoskeleton is composed of microfilaments, intermediate filaments and microtubules. There is a great number of proteins associated with them, each controlling a cell's structure by directing, bundling, and aligning filaments. The prokaryotic cytoskeleton is less well-studied but is involved in the maintenance of cell shape, polarity and cytokinesis.

Organelles

The human body contains many different organs, such as the heart, lung, and kidney, with each organ performing a different function. Cells also have a set of "little organs," called organelles, that are adapted and/or specialized for carrying out one or more vital functions. Both eukaryotic and prokaryotic cells have organelles but organelles in eukaryotes are generally more complex and may be membrane bound.

There are several types of organelles in a cell. Some (such as the nucleus and golgi apparatus) are typically solitary, while others (such as mitochondria, peroxisomes and lysosomes) can be numerous (hundreds to thousands). The cytosol is the gelatinous fluid that fills the cell and surrounds the organelles.

Structures Outside the Cell Wall

Capsule

A gelatinous capsule is present in some bacteria outside the cell wall. The capsule may be polysaccharide as in pneumococci, meningococci or polypeptide as *Bacillus anthracis* or hyaluronic acid as in streptococci. Capsules are not marked by ordinary stain and can be detected by special stain. The capsule is antigenic. The capsule has antiphagocytic function so it determines the virulence of many bacteria. It also plays a role in attachment of the organism to mucous membranes.

Flagella

Flagella are the organelles of cellular mobility. They arise from cytoplasm and extrude through the cell wall. They are long and thick thread-like appendages, protein in nature. Are most commonly found in bacteria cells but are found in animal cells as well.

Fimbriae (Pili)

They are short and thin hair like filaments, formed of protein called pilin (antigenic). Fimbriae are responsible for attachment of bacteria to specific receptors of human cell (adherence). There are special types of pili called (sex pili) involved in conjunction.

Functions

Growth and Metabolism

Between successive cell divisions, cells grow through the functioning of cellular metabolism. Cell metabolism is the process by which individual cells process nutrient molecules. Metabolism has two distinct divisions: catabolism, in which the cell breaks down complex molecules to produce energy and reducing power, and anabolism, in which the cell uses energy and reducing power to construct complex molecules and perform other biological functions. Complex sugars consumed by the organism can be broken down into a less chemically complex sugar molecule called glucose. Once inside the cell, glucose is broken down to make adenosine triphosphate (ATP), a form of energy, through two different pathways.

The first pathway, glycolysis, requires no oxygen and is referred to as anaerobic metabolism. Each reaction is designed to produce some hydrogen ions that can then be used to make energy packets (ATP). In prokaryotes, glycolysis is the only method used for converting energy.

The second pathway, called the Krebs cycle, or citric acid cycle, occurs inside the mitochondria and can generate enough ATP to run all the cell functions.

Creation

Cell division involves a single cell (called a *mother cell*) dividing into two daughter cells. This leads to growth in multicellular organisms (the growth of tissue) and to procreation (vegetative reproduction) in unicellular organisms.

Prokaryotic cells divide by binary fission. Eukaryotic cells usually undergo a process of nuclear division, called mitosis, followed by division of the cell, called cytokinesis. A diploid cell may also undergo meiosis to produce haploid cells, usually four. Haploid cells serve as gametes in multicellular organisms, fusing to form new diploid cells.

DNA replication, or the process of duplicating a cell's genome, is required every time a cell divides. Replication, like all cellular activities, requires specialized proteins for carrying out the job.

Protein Synthesis

Cells are capable of synthesizing new proteins, which are essential for the modulation and maintenance of cellular activities. This process involves the formation of new protein molecules from amino acid building blocks based on information encoded in DNA/RNA. Protein synthesis generally consists of two major steps: transcription and translation. Transcription is the process where genetic information in DNA is used to produce a complementary RNA strand. This RNA strand is then processed to give messenger RNA (mRNA), which is free to migrate through the cell. mRNA molecules bind to protein-RNA complexes called ribosomes located in the cytosol, where they are translated into polypeptide sequences. The ribosome mediates the formation of a polypeptide sequence based on the mRNA sequence. The mRNA sequence directly relates to the polypeptide sequence by binding to transfer RNA (tRNA) adapter molecules in binding pockets within the ribosome. The new polypeptide then folds into a functional three-dimensional protein molecule.

Movement or Motility

Cells can move during many processes: such as wound healing, the immune response and cancer metastasis. For wound healing to occur, white blood cells and cells that ingest bacteria move to the wound site to kill the microorganisms that cause infection.

At the same time fibroblasts (connective tissue cells) move there to remodel damaged structures. In the case of tumour development, cells from a primary tumour move away and spread to other parts of the body. Cell motility involves many receptors, crosslinking, bundling, binding, adhesion, motor and other proteins. The process is divided into three steps – protrusion of the leading edge of the cell, adhesion of the leading edge and de-adhesion at the cell body and rear, and cytoskeletal contraction to pull the cell forward. Each step is driven by physical forces generated by unique segments of the cytoskeleton.

Evolution

The origin of cells has to do with the origin of life, which began the history of life on Earth.

Origin of the First Cell

There are several theories about the origin of small molecules that could lead to life in an early Earth. One is that they came from meteorites. Another is that they were created at deep-sea vents. A third is that they were synthesized by lightning in a reducing atmosphere; although it is not clear if Earth had such an atmosphere. There are essentially no experimental data defining what the first self-replicating forms were. RNA is generally assumed to be the earliest self-replicating molecule, as it is capable of both storing genetic information and catalysing chemical reactions. But some other entity with the potential to self-replicate could have preceded RNA, like clay or peptide nucleic acid.

Cells emerged at least 4.0–4.3 billion years ago. The current belief is that these cells were heterotrophs. An important characteristic of cells is the cell membrane, composed of a bilayer of lipids. The early cell membranes were probably more simple and permeable than modern ones, with only a single fatty acid chain per lipid. Lipids are known to spontaneously form bilayered vesicles in water, and could have preceded RNA. But the first cell membranes could also have been produced by catalytic RNA, or even have required structural proteins before they could form.

Origin of Eukaryotic Cells

The eukaryotic cell seems to have evolved from a symbiotic community of prokaryotic cells. DNA-bearing organelles like the mitochondria and the chloroplasts are almost certainly what remains of ancient symbiotic oxygen-breathing proteobacteria and cyanobacteria, respectively, where the rest of the cell seems to be derived from an ancestral archaean prokaryote cell – a theory termed the endosymbiotic theory.

There is still considerable debate about whether organelles like the hydrogenosome predated the origin of mitochondria.

Sex, as the stereotyped choreography of meiosis and syngamy that persists in nearly all extant eukaryotes, may have played a role in the transition from prokaryotes to eukaryotes.

An 'origin of sex as vaccination' theory suggests that the eukaryote genome accreted from prokaryan parasite genomes in numerous rounds of lateral gene transfer. Sex-as-syngamy (fusion sex) arose when infected hosts began swapping nuclearized genomes containing co-evolved, vertically transmitted symbionts that conveyed protection against horizontal infection by more virulent symbionts.

History

- 1632–1723: Antonie van Leeuwenhoek teaches himself to grind lenses, builds a microscope and draws protozoa, such as *Vorticella* from rain water, and bacteria from his own mouth.
- 1665: Robert Hooke discovers cells in cork, then in living plant tissue using an early microscope.
- 1839: Theodor Schwann and Matthias Jakob Schleiden elucidate the principle that plants and animals are made of cells, concluding that cells are a common unit of structure and development, and thus founding the cell theory.
- The belief that life forms can occur spontaneously (*generatio spontanea*) is contradicted by Louis Pasteur (1822–1895)

(although Francesco Redi had performed an experiment in 1668 that suggested the same conclusion).

- 1855: Rudolph Virchow states that cells always emerge from cell divisions (*omnis cellula ex cellula*).

- 1931: Ernst Ruska builds first transmission electron microscope (TEM) at the University of Berlin. By 1935, he has built an EM with twice the resolution of a light microscope, revealing previously unresolvable organelles.

- 1953: Watson and Crick made their first announcement on the double-helix structure for DNA on February 28.

- 1981: Lynn Margulis published *Symbiosis in Cell Evolution* detailing the endosymbiotic theory.

5

Mechanical Separation
of Nucleic Acids

Isolation of Nucleic Acids

Isolation of nucleic acid is performed in the laboratory for a variety of reasons such as cloning of desired genes, comparisons of different genomes, study of expression patterns, and forensic evaluation. No matter the reason for the isolation of the nucleic acid, the general methods are similar.

Methods

1. Lysing the cell. One must first break open the cell to release the nucleic acids.

    a. Detergents-dissolve the lipid membrane of cells

    b. Enzymatic digestion to remove the cell wall of Gram + bacteria and plant cells. Examples-mutanolysin and lysostaphin

    c. Physical disruption-grinding or homogenizing to remove cell wall

2. Remove contaminating material from nucleic acids

    a. Enzymatic digestion-proteinases digest proteins

    b. Organic solvent extraction-proteins, but not nucleic acids, dissolve in phenol and chloroform.

    c. Chromatographic methods-Anionic columns (recall that DNA is negatively charged; therefore, it binds to positively charged substances), glass or silica bases methods. DNA, but not other components, binds to the material of choice. Once the contaminating components are washed away, the DNA is eluted from the binding matrix.

3. Purification/concentration of nucleic acid

   a. Precipitation with alcohol-DNA is insoluble in either ethanol or isopropanol.

   b. Centrifugation-centrifugal force causes DNA to form pellet in bottom of tube.

   c. Dialysis-concentrates and cleans DNA by removing salts and other impurities that are small enough to migrate through dialysis membrane.

4. Differentiation of nucleic acids

   Genomic DNA and plasmid DNA

   a. Cesium chloride gradients-plasmid DNA and genomic DNA differ in density. Genomic DNA is less dense than plasmid DNA.

   b. Alkaline lysis-Alkaline conditions denature DNA. Upon neutralization, plasmid DNA is capable of renaturing whereas genomic DNA remains denatured.

Protocols for Isolating DNA

Genomic DNA Isolation

Cesium chloride/Ethidium bromide Gradients for Extraction of Chromosomal DNA

1. Spin down cells (5K rpm 5-10 min).

2. Remove supernatant and freeze pellet at -80°C for at least 15 min. (If concerned about nucleases, resuspend pellet in 1X TE buffer to remove excess media and spin at 14K rpm in SA600 or SS34 rotor 5 min and skip deep freeze of pellet.)

3. Resuspend pellet in 2.5 ml of cold 25% sucrose in 50 mM Tris, pH 8.0.

4. Add 0.5 ml lysozyme (10 mg/ml in 0.25 M Tris, pH 8.0). Incubate on ice 10 min.

5. Add 1 ml cold 0.25 M EDTA, pH 8.0 and incubate on ice 10 min.

6. Add 4 ml cold Triton X-100 lysing solution (10% triton X-100, 50 mM Tris, pH 8.0, 100 mM EDTA, pH 8.0) and mix gently. Sample should become viscous indicating cell lysis.

7. Collect supernatant in graduated cylinder and add buffer to 11 ml.

8. Add 11g CsCl and 1 ml EtBr to supernatant.

9. Fill, balance, and seal tubes. Spin at 45K rpm for 44 hr at 20°C in Ti75 rotor in Beckman ultracentrifuge.

10. Place the tubes in front of UV light source to visualize the DNA. (Use a full face shield to protect yourself from the UV light).

11. Use a needle and syringe to remove the chromosomal DNA band from the sealed tubes.

12. Remove the EtBr with isopropanol saturated with water and CsCl.

13. Concentrate and remove the CsCl from the DNA by dialysis against TE buffer using a Schleicher and Schuell rapid dialysis chamber.

Various commercially available kits may also be used to isolate genomic DNA. Some suppliers are listed below.

- Promega
- QIAGEN
- Eppendorf Scientific.

Plasmid DNA Isolation

A. Boiling Lysis Miniprep DNA Isolation (From Maniatis)

1. Inoculate 5 ml of LB medium supplemented with the appropriate antibiotic with a single colony from a plate culture. Grow overnight at 37°C with shaking.

2. The next morning, transfer 1.5 ml of culture to eppendorf tube and pellet cells by centrifugation for 2 min at 14K rpm in a microcentrifuge.

3. Resuspend pellet in 350 μl STET (100 mM NaCl, 100 mM Tris (pH 8.0), 1 mM EDTA, 5% Triton X-100).

4. Add 25 μl lysozyme (10 mg/ml in 0.25 M Tris, pH 8.0). Briefly vortex to mix.

5. Place samples in boiling water bath for 40 sec.

6. Pellet cell debris (membrane, chromosomal DNA, proteins, etc...) by centrifugation at 14K rpm for 10 min at room temperature.

7. Remove pellet with a sterile toothpick.

8. Precipitate DNA by adding 33.3 μl 3M sodium acetate, pH 5.2 and 420 μl isopropanol. Incubate at room temperature for 5 min.

9. Pellet DNA by centrifugation at 14K rpm at 4°C for 10 min.

10. Wash pellet with 70% ethanol.

11. Dry pellet in speed vac or in desiccator for 5 min.

12. Resuspend pellet in 50 µl TE containing 20 µg/ml RNase. Store sample at -20°C.

B. Alkaline Lysis Method (taken from ABI Technologies Instruction Manual).

This method results in DNA that is sufficiently clean to use for PCR or sequencing.

1. Inoculate 5 ml of LB medium supplemented with the appropriate antibiotic with a single colony from a plate culture. Grow overnight at 37°C with shaking.

2. Pellet the entire culture in a single microfuge tube.

3. Resuspend the pellet was in 200 µl GTE buffer (50 mM glucose, 25 mM Tris, 10 mM EDTA; pH 8.0).

4. Add 300 µl of freshly prepared 0.2 N NaOH, 1% SDS, mix the by inversion and incubate on ice 5 min.

5. Add 300 µl 3 M KOAc (pH 4.8), mix the samples inversion and incubate on ice 5 min.

6. Centrifuge for 10 min at room temperature.

7. Transfer the supernatant to a new tube.

8. Add RNase to a final concentration of 20 µg/ml and incubate the samples for 20 min at 37°C.

9. Extract the samples two times with 400 µl of chloroform. (Mix layers gently by hand to avoid shearing chromosomal DNA.

10. Precipitate total DNA by adding an equal volume of 100% isopropanol.

11. Immediately centrifuge 10 min at room temperature.

12. Wash the pellet with 500 µl 70% ethanol.

13. Dry under vacuum for 3 min.

14. Resuspend pellet in 32 µl of sterile H2O, 8 µl 4 M NaCl, and 40 µl 13% PEG8000 and incubate on ice 20 min.

15. Centrifuge for 15 min at 4°C.

16. Wash the plasmid pellet with 500 µl 70% ethanol and dry under vacuum for 3 min.

17. Resuspend the final pellet in 20 μl sterile H2O and store at -20°C.

C. Large scale DNA preparation (CsCl gradient)

1. Inoculate LB medium (250 ml for high copy plasmids or 500 ml for low copy plasmids) containing the appropriate antibiotic with 5 ml of an overnight starter culture of *E. coli* harboring the desired plasmid.

2. Incubate the culture at 37°C with shaking until the OD600 reaches 0.4.

3. Add chloramphenicol to a final concentration of 170 μg/ml and continue the 12-16 h.

4. Harvest the cells by centrifugation at 5000 rpm for 10 min at 4°C.

5. Incubate the cell at -80°C for 15 min.

6. After thawing, resuspend the cell pellet in 2.5 ml of cold 25% sucrose in 50 mM Tris, pH 8.0.

7. Add 0.5 ml of Lysozyme (10 mg/ml solution in 0.25 M Tris, pH 8.0).

8. Incubate on ice for 10 min.

9. Add 1.0 ml of 0.25 M EDTA (pH 8.0) and continue the incubation for 10 min.

10. Lyse cells by adding 4.0 ml of cold Triton X-100 lysing solution (10% Triton X-100, 50 mM Tris, 62.5 mM EDTA, pH 8.0) and mixing gently by inversion.

11. Centrifufe the lysate at 15,000 rpm for 15 min at 4°C in a SS-34 rotor to pellet the cell debris and chromosomal DNA.

12. Collect the supernatant and bring to a final volume of 10 ml with TE buffer (10 mM Tris, 1 mM EDTA, pH 8.0).

13. Add CsCl and EtBr to a final concentration of 1 g/ml and 1 mg/ml, respectively.

14. Place the sample was in quick seal tubes (Beckman, Columbia, MD) and centrifuge in a Ti75 rotor at 45,000 rpm for 44 h at 25°C.

15. Place the tubes in front of UV light source to visualize the DNA. (Use a full face shield to protect yourself from the UV light).

16. Use a needle and syringe to remove the plasmid DNA band from the sealed tubes.

17. Remove the EtBr with isopropanol saturated with water and CsCl.

18. Concentrate and remove the CsCl from the DNA by dialysis against TE buffer using a Schleicher and Schuell rapid dialysis chamber.

Various commercially available kits may also be used to isolate plasmid DNA. Several suppliers are listed below.

- QIAGEN
- Biorad
- Life Technologies
- Bio 101
- Promega
- Eppendorf Scientific
- Invitrogen.

Quantitation of Nucleic Acids

A. Spectrophotometric Quantitation:

1. Turn spectrophotometer on and allow the UV light source to warm up.

2. Dilute nucleic acid sample in water or TE buffer so that the final volume of the sample will be 1 ml.

3. Prepare a blank that contains only the diluant (water or TE buffer).

4. Place blank solution into cuvette (either quartz or methylacrylate when taking readings in the UV range) and calibrate machine.

5. Place sample into cuvette and take readings at both 260 nm and 280 nm.

6. Calculate nucleic acid concentrations according to the following equations:

[dsDNA in $\mu g/\mu l$] = A_{260} x 50 μg

volume of DNA added to cuvette

[ssDNA or RNA] = A_{260} x 40 μg

volume of nucleic acid added to cuvette.

DNA Methylation Analysis

DNA analysis is an examination method that emerged in the 1980s and is credited to Alec Jeffreys, an English geneticist. Every species has unique genetic sequences. DNA analysis allows any type of organism to be identified by analysing its genetic sequences. This method can also clarify questions of identification within a species.

Identification within a species can present more of a challenge than determining between two different species. For example, it is much easier to determine whether a victim was attacked by a bear or human than it is to determine which human perpetrated an attack. DNA analysis is usually performed by forensic scientists. In instances where individuals need to be identified, the forensic scientists tend to employ a method of scanning 13 regions of the human genome.

If forensic scientists only scanned a few areas, DNA analysis would not likely be highly regarded as accurate. The chances of an identical match between two people when 13 regions are scanned are so slim that the results are sometimes called a DNA fingerprint. Like the print left from an index finger, DNA fingerprints are generally considered conclusive and are not usually subject to much scrutiny.

There are several techniques that can be used for DNA analysis. Restriction fragment length polymorphism (RFLP) was one of the first methods used in forensic investigation. Polymerase chain reaction (PCR) is a method that allows a forensic scientist to amplify a sample and make millions of copies of the DNA from a relatively small specimen.

One factor that determines the method a forensic scientist will employ is how big her specimen is. An entire body is not necessary to perform a DNA analysis, and oftentimes a large specimen is not available. One hair, a tooth, a drop of blood, or skin cells usually contain enough unique information about a person to identify him when the proper technique is used.

The question that needs to be answered also determines the method that is used for DNA analysis. Every father passes his Y chromosome to his son. A Y chromosome analysis therefore can be employed in situations where the issue is establishing paternity.

Since DNA analysis is respected as being highly accurate, it is used in many parts of the world for many important reasons. Examples include to identify individuals who are suspected criminals, to identify victims or their remains when identity by other means is not possible, to determine the risk of genetic diseases, and to identify animals in cases of suspected poaching. A person can spend his life in prison or be freed from a sentence of life based on the results from DNA analysis.

Types of DNA Analysis

PCR Analysis

PCR (polymerase chain reaction) enhances DNA analysis and has enabled laboratories to develop DNA profiles from extremely small samples of biological evidence. The PCR technique replicates exact copies of DNA contained in a biological evidence sample without affecting the original, much like a copy machine. RFLP analysis requires a biological sample about the size of a quarter, but PCR can be used to reproduce millions of copies of the DNA contained in a few skin cells.

Since PCR analysis requires only a minute quantity of DNA, it can enable the laboratory to analyze highly degraded evidence for DNA. On the other hand, because the sensitive PCR technique replicates any and all of the DNA contained in an evidence sample, greater attention to contamination issues is necessary when identifying, collecting, and preserving DNA evidence. These factors may be particularly important in the evaluation of unsolved cases in which evidence might have been improperly collected or stored.

STR Analysis

Short tandem repeat (STR) technology is a forensic analysis that evaluates specific regions (loci) that are found on nuclear DNA. The variable (polymorphic) nature of the STR regions that are analysed for forensic testing intensifies the discrimination between one DNA profile and another. For example, the likelihood that any two individuals (except identical twins) will have the same 13-loci DNA profile can be as high as 1 in 1 billion or greater. The Federal Bureau of Investigation (FBI) has chosen 13 specific STR loci to serve as the standard for CODIS. The purpose of establishing a core set of STR loci is to ensure that all forensic laboratories can establish uniform DNA databases and, more importantly, share valuable forensic information. If the forensic or convicted offender CODIS index is to be used in the investigative stages of unsolved cases, DNA profiles must be generated by using STR technology and the specific 13 core STR loci selected by the FBI.

Mitochondrial DNA Analysis

Mitochondrial DNA (mtDNA) analysis allows forensic laboratories to develop DNA profiles from evidence that may not be suitable for RFLP or STR analysis. While RFLP and PCR techniques analyze DNA extracted from the nucleus of a cell, mtDNA technology analyzes DNA found in a different part of the cell, the mitochondrion. Old remains and evidence lacking nucleated cells—such as hair shafts, bones, and teeth—that are unamenable to STR and RFLP testing may yield results if mtDNA analysis is performed.

For this reason, mtDNA testing can be very valuable to the investigation of an unsolved case. For example, a cold case log may show that biological evidence in the form of blood, semen, and hair was collected in a particular case, but that all were improperly stored for a long period of time. Although PCR analysis sometimes enables the crime laboratory to generate a DNA profile from very degraded evidence, it is possible that the blood and semen would be so highly degraded that nuclear DNA analysis would not yield a DNA profile. However, the hair shaft could be subjected to mtDNA analysis and thus be the key to solving the case.

Finally, it is important to note that all maternal relatives (for example, a person's mother or maternal grandmother) have identical mtDNA. This enables unidentified remains to be analysed and compared to the mtDNA profile of any maternal relative for the purpose of aiding missing persons or unidentified remains investigations. Although mtDNA analysis can be very valuable to the investigation of criminal cases, laboratory personnel should always be involved in the process.

Y-chromosome Analysis

Y-chromosome analysis Several genetic markers have been identified on the Y chromosome that can be used in forensic applications. Ychromosome markers target only the male fraction of a biological sample. Therefore, this technique can be very valuable if the laboratory detects complex mixtures (multiple male contributors) within a biological evidence sample. Because the Y chromosome is transmitted directly from a father to all of his sons, it can also be used to trace family relationships among males. Advancements in Y-chromosome testing may eventually eliminate the need for laboratories to extract and separate semen and vaginal cells (for example, from a vaginal swab of a rape kit) prior to analysis. Cooperative efforts with the crime laboratory are essential to deciding which analysis methods will be most valuable in a particular case. It is important to note, however, that while RFLP and mtDNA testing may be valuable to the investigation of an old case, current DNA databases are being populated with DNA profiles that are generated using STR analysis.

RFLP and mtDNA profiles are not compatible with the convicted offender or forensic indexes of CODIS. (CODIS has a missing persons index that exclusively contains mtDNA profiles; the convicted offender and forensic indexes of CODIS exclusively contain STR DNA profiles.)

DNA methylation is an epigenetic event that affects cell function by altering gene expression and refers to the covalent addition of a methyl group, catalysed by DNA methyltransferase (DNMT), to the 5-carbon of cytosine in a CpG dinucleotide. Methods for DNA methylation analysis can be divided roughly into two types: global and gene-specific methylation analysis. For global methylation analysis, there are methods which measure the overall level of methyl cytosines in genome such as chromatographic methods and methyl accepting capacity assay.

For gene-specific methylation analysis, a large number of techniques have been developed. Most early studies used methylation sensitive restriction enzymes to digest DNA followed by Southern detection or PCR amplification. Recently, bisulfite reaction based methods have become very popular such as methylation specific PCR (MSP), bisulfite genomic sequencing PCR. Additionally, in order to identify unknown methylation hot-spots or methylated CpG islands in the genome, several of genome-wide screen methods have been invented such as Restriction Landmark Genomic Scanning for Methylation (RLGS-M), and CpG island microarray.

Methylation

In the chemical sciences, methylation denotes the addition of a methyl group to a substrate or the substitution of an atom or group by a methyl group. Methylation is a form of alkylation with, to be specific, a methyl group, rather than a larger carbon chain, replacing a hydrogen atom. These terms are commonly used in chemistry, biochemistry, soil science, and the biological sciences.

In biological systems, methylation is catalysed by enzymes; such methylation can be involved in modification of heavy metals, regulation of gene expression, regulation of protein function, and RNA metabolism.

Methylation of heavy metals can also occur outside of biological systems. Chemical methylation of tissue samples is also one method for reducing certain histological staining artifacts.

Biological Methylation

Epigenetics

Methylation contributing to epigenetic inheritance can occur through either DNA methylation or protein methylation.

DNA methylation in vertebrates typically occurs at CpG sites (cytosine-phosphate-guanine sites, that is, where a cytosine is directly followed by a guanine in the DNA sequence). This methylation results in the conversion of the cytosine to 5-methylcytosine. The formation of Me-CpG is catalysed by the enzyme DNA methyltransferase. Human DNA has about 80%-90% of CpG sites methylated, but there are certain areas, known as CpG islands, that are GC-rich (made up of about 65% CG residues), wherein none are methylated. These are associated with the promoters of 56% of mammalian genes, including all ubiquitously expressed genes. One to two percent of the human genome are CpG clusters, and there is an inverse relationship between CpG methylation and transcriptional activity.

Protein methylation typically takes place on arginine or lysine amino acid residues in the protein sequence. Arginine can be methylated once (monomethylated arginine) or twice, with either both methyl groups on one terminal nitrogen (asymmetric dimethylated arginine) or one on both nitrogens (symmetric dimethylated arginine) by peptidylarginine methyltransferases (PRMTs). Lysine can be methylated once, twice or three times by lysine methyltransferases. Protein methylation has been most-studied in the histones. The transfer of methyl groups from S-adenosyl methionine to histones is catalysed by enzymes known as histone methyltransferases. Histones that are methylated on certain residues can act epigenetically to repress or activate gene expression. Protein methylation is one type of post-translational modification.

Embryonic Development

In early mammalian development, the genome within the germ cells is demethylated, while chromosomes in the remaining cells retain the parental methylation patterns. *De novo* methylation of the germ cells occurs, modifying and adding epigenetic information to the genome based on the sex of the individual. By blastula stage, the methylation is complete. This process is referred to as "reprogramming". The importance of methylation was shown in knockout mutants without DNA methyltransferase, which all died at the morula stage.

Postnatal Development

Increasing evidence is revealing a role of methylation in the interaction of environmental factors with genetic expression. Differences in maternal care during the first 6 days of life in the rat induce differential methylation patterns in some promoter regions and, thus, influencing gene expression. Furthermore, even-more-dynamic processes such as interleukin signalling have been shown to be regulated by methylation.

Cancer

The pattern of methylation has recently become an important topic for research. Studies have found that in normal tissue, methylation of a gene is mainly localised to the coding region, which is CpG-poor. In contrast, the promoter region of the gene is unmethylated, despite a high density of CpG islands in the region.

Neoplasia is characterized by "methylation imbalance" where genome-wide hypomethylation is accompanied by localized hypermethylation and an increase in expression of DNA methyltransferase.

The overall methylation state in a cell might also be a precipitating factor in carcinogenesis as evidence suggests that genome-wide hypomethylation can lead to chromosome instability and increased mutation rates. The methylation state of some genes can be used as a biomarker for tumorigenesis. For instance, hypermethylation of the pi-class glutathione S-transferase gene (GSTP1) appears to be a promising diagnostic indicator of prostate cancer.

In cancer, the dynamics of genetic and epigenetic gene silencing are very different.

Somatic genetic mutation leads to a block in the production of functional protein from the mutant allele. If a selective advantage is conferred to the cell, the cells expand clonally to give rise to a tumour in which all cells lack the capacity to produce protein.

In contrast, epigenetically mediated gene silencing occurs gradually. It begins with a subtle decrease in transcription, fostering a decrease in protection of the CpG island from the spread of flanking heterochromatin and methylation into the island. This loss results in gradual increases of individual CpG sites, which vary between copies of the same gene in different cells.

Bacterial Host Defence

In addition, adenosine or cytosine methylation is part of the restriction modification system of many bacteria. Bacterial DNAs are methylated periodically throughout the genome. A methylase is the enzyme that recognizes a specific sequence and methylates one of the bases in or near that sequence.

Foreign DNAs (which are not methylated in this manner) that are introduced into the cell are degraded by sequence-specific restriction enzymes.

Bacterial genomic DNA is not recognized by these restriction enzymes. The methylation of native DNA acts as a sort of primitive immune system, allowing the bacteria to protect themselves from infection by bacteriophage. These restriction enzymes are the basis of restriction fragment length polymorphism (RFLP) testing, used to detect DNA polymorphisms.

Methylation in Chemistry

The term methylation in organic chemistry refers to the alkylation process used to describe the delivery of a CH_3 group.

This is commonly performed using *electrophilic* methyl sources - iodomethane, dimethyl sulfate, dimethyl carbonate, or less commonly with the more powerful (and more dangerous) methylating reagents of methyl triflate or methyl fluorosulfonate (magic methyl), which all react via S_N2 nucleophilic substitution. For example a carboxylate may be methylated on oxygen to give a methyl ester, an alkoxide salt RO^- may be likewise methylated to give an ether, $ROCH_3$, or a ketone enolate may be methylated on carbon to produce a new ketone.

On the other hand, the methylation may involve use of *nucleophilic* methyl compounds such as methyllithium (CH_3Li) or Grignard reagents (CH_3MgX). For example, CH_3Li will methylate acetone, adding across the carbonyl (C=O) to give the lithium alkoxide of *tert*-butanol:

Purdie Methylation

Purdie methylation is a specific method for the methylation at oxygen of carbohydrates using iodomethane and silver oxide

DNA Methylation

DNA methylation involves the addition of a methyl group to the 5 position of the cytosine pyrimidine ring or the number 6 nitrogen of the adenine purine ring (cytosine and adenine are two of the four bases of DNA). This modification can be inherited through cell division. DNA methylation is typically removed during zygote formation and re-established through successive cell divisions during development. DNA methylation is a crucial part of normal organismal development and cellular differentiation in higher organisms.

DNA methylation stably alters the gene expression pattern in cells such that cells can "remember where they have been"; in other words, cells programmed to be pancreatic islets during embryonic development remain pancreatic islets throughout the life of the organism without continuing signals telling them that they need to remain islets. In addition, DNA methylation suppresses the expression of viral genes and other deleterious elements that have been incorporated into the genome of the host over time. DNA methylation also forms the basis of chromatin structure, which enables cells to form the myriad characteristics necessary for multicellular life from a single immutable sequence of DNA. DNA methylation also plays a crucial role in the development of nearly all types of cancer.

DNA methylation involves the addition of a methyl group to DNA — for example, to the number 5 carbon of the cytosine pyrimidine ring — in this case with the specific effect of reducing gene expression. DNA methylation at the 5 position of cytosine has been found in every vertebrate examined. In adult somatic tissues, DNA methylation typically occurs in a CpG dinucleotide context; non-CpG methylation is prevalent in embryonic stem cells.

In Mammals

DNA methylation is essential for normal development and is associated with a number of key processes including genomic imprinting, X-chromosome inactivation, suppression of repetitive elements, and carcinogenesis.

Between 60% and 90% of all CpGs are methylated in mammals. Methylated C residues spontaneously deaminate to form T residues; hence CpG dinucleotides steadily mutate to TpG dinucleotides, which is evidenced by the under-representation of CpG dinucleotides in the human genome (they occur at only 21% of the expected frequency). (On the other hand, spontaneous deamination of unmethylated C residues gives rise to U residues, a mutation that is quickly recognized and repaired by the cell)

Unmethylated CpGs are often grouped in clusters called *CpG islands*, which are present in the 5' regulatory regions of many genes. In many disease processes such as cancer, gene promoter CpG islands acquire abnormal hypermethylation, which results in transcriptional silencing that can be inherited by daughter cells following cell division. Alterations of DNA methylation have been recognized as an important component of cancer development. Hypomethylation, in general, arises earlier and is linked to chromosomal instability and loss of imprinting, whereas hypermethylation is associated with promoters and can arise secondary to gene (oncogene suppressor) silencing, but might be a target for epigenetic therapy.

DNA methylation may affect the transcription of genes in two ways. First, the methylation of DNA may itself may physically impede the binding of transcriptional proteins to the gene, and second, and likely more important, methylated DNA may be bound by proteins known as methyl-CpG-binding domain proteins (MBDs). MBD proteins then recruit additional proteins to the locus, such as histone deacetylases and other chromatin remodeling proteins that can modify histones, thereby forming compact, inactive chromatin, termed silent chromatin. This link between DNA methylation and chromatin structure is very important. In particular, loss of methyl-CpG-binding protein 2 (MeCP2) has been implicated in Rett syndrome; and methyl-CpG-binding domain protein 2 (MBD2) mediates the transcriptional silencing of hypermethylated genes in cancer.

Research has suggested that long-term memory storage in humans may be regulated by DNA methylation.

Epigenetics

Epigenetics is the study of inherited changes in phenotype (appearance) or gene expression caused by mechanisms other than changes in the underlying DNA sequence, hence the name *epi-* (- over, above) -*genetics*. These changes may remain through cell divisions for the remainder of the cell's life and may also last for multiple generations. However, there is no change in the underlying DNA sequence of the organism; instead, non-genetic factors cause the organism's genes to behave (or "express themselves") differently.

The best example of epigenetic changes in eukaryotic biology is the process of cellular differentiation. During morphogenesis, totipotent stem cells become the various pluripotent cell lines of the embryo which in turn become fully differentiated cells. In other words, a single fertilized egg cell – the zygote – changes into the many cell types including neurons, muscle cells, epithelium, blood vessels etc. as it continues to divide. It does so by activating some genes while inhibiting others.

Etymology and Definitions

Epigenetics (as in "epigenetic landscape") was coined by C. H. Waddington in 1942 as a portmanteau of the words *genetics* and *epigenesis*. *Epigenesis* is an old word which has more recently been used to describe the differentiation of cells from their initial totipotent state in embryonic development. When Waddington coined the term the physical nature of genes and their role in heredity was not known; he used it as a conceptual model of how genes might interact with their surroundings to produce a phenotype.

Robin Holliday defined epigenetics as "the study of the mechanisms of temporal and spatial control of gene activity during the development of complex organisms." Thus *epigenetic* can be used to describe anything other than DNA sequence that influences the development of an organism.

The modern usage of the word in scientific discourse is more narrow, referring to heritable traits (over rounds of cell division and sometimes transgenerationally) that do not involve changes to the underlying DNA sequence. The Greek prefix *epi-* in *epigenetics* implies features that are "on top of" or "in addition to" genetics; thus *epigenetic* traits exist on top of or in addition to the traditional molecular basis for inheritance.

The similarity of the word to "genetics" has generated many parallel usages. The "epigenome" is a parallel to the word "genome", and refers to the overall epigenetic state of a cell. The phrase "genetic code" has also been adapted—the "epigenetic code" has been used to describe the set of epigenetic features that create different phenotypes in different cells. Taken to its extreme, the "epigenetic code" could represent the total state of the cell, with the position of each molecule accounted for in an *epigenomic map*, a diagrammatic representation of the gene expression, DNA methylation and histone modification status of a particular genomic region. More typically, the term is used in reference to systematic efforts to measure specific, relevant forms of epigenetic information such as the histone code or DNA methylation patterns.

The psychologist Erik Erikson used the term *epigenetic* in his theory of psychosocial development. That usage, however, is of primarily historical interest.

Molecular Basis of Epigenetics

The molecular basis of epigenetics is complex. It involves modifications of the activation of certain genes, but not the basic structure of DNA. Additionally, the chromatin proteins associated with DNA may be activated or silenced. This accounts for why the differentiated cells in a multi-cellular organism express only the genes that are necessary for their own activity. Epigenetic changes are preserved when cells divide. Most epigenetic changes only occur within the course of one individual organism's lifetime, but, if a mutation in the DNA has been caused in sperm or egg cell that results in fertilization, then some epigenetic changes are inherited from one generation to the next. This raises the question of whether or not epigenetic changes in an organism can alter the basic structure of its DNA, a form of Lamarckism.

Specific epigenetic processes include paramutation, bookmarking, imprinting, gene silencing, X chromosome inactivation, position effect, reprogramming, transvection, maternal effects, the progress of carcinogenesis, many effects of teratogens, regulation of histone modifications and heterochromatin, and technical limitations affecting parthenogenesis and cloning.

Epigenetic research uses a wide range of molecular biologic techniques to further our understanding of epigenetic phenomena, including chromatin immunoprecipitation (together with its large-scale variants ChIP-on-chip and ChIP-seq), fluorescent in situ hybridization, methylation-sensitive restriction enzymes, DNA adenine methyltransferase identification (DamID) and bisulfite sequencing. Furthermore, the use of bioinformatic methods is playing an increasing role (computational epigenetics).

Mechanisms

Several types of epigenetic inheritance systems may play a role in what has become known as cell memory:

DNA Methylation and Chromatin Remodeling

Because the phenotype of a cell or individual is affected by which of its genes are transcribed, heritable transcription states can give rise to epigenetic effects. There are several layers of regulation of gene expression. One way that genes are regulated is through the remodeling of chromatin. Chromatin is the complex of DNA and the histone proteins with which it associates. Histone proteins are little spheres that DNA wraps around. If the way that DNA is wrapped around the histones changes, gene expression can change as well. Chromatin remodeling is accomplished through two main mechanisms:

1. The first way is post translational modification of the amino acids that make up histone proteins. Histone proteins are made up of long chains of amino acids. If the amino acids that are in the chain are changed, the shape of the histone sphere might be modified. DNA is not completely unwound during replication. It is possible, then, that the modified histones may be carried into each new copy of the DNA. Once there, these histones may act as templates, initiating the surrounding new histones to be shaped in the new manner. By altering the shape of the histones around it, these modified histones would ensure that a differentiated cell would stay differentiated, and not convert back into being a stem cell.

2. The second way is the addition of methyl groups to the DNA, mostly at CpG sites, to convert cytosine to 5-methylcytosine. 5-Methylcytosine performs much like a regular cytosine, pairing up with a guanine. However, some areas of genome are methylated more heavily than others and highly methylated

areas tend to be less transcriptionally active, through a mechanism not fully understood. Methylation of cytosines can also persist from the germ line of one of the parents into the zygote, marking the chromosome as being inherited from this parent (genetic imprinting).

The way that the cells stay differentiated in the case of DNA methylation is clearer to us than it is in the case of histone shape. Basically, certain enzymes (such as DNMT1) have a higher affinity for the methylated cytosine. If this enzyme reaches a "hemimethylated" portion of DNA (where methylcytosine is in only one of the two DNA strands) the enzyme will methylate the other half.

Although histone modifications occur throughout the entire sequence, the unstructured N-termini of histones (called histone tails) are particularly highly modified. These modifications include acetylation, methylation, ubiquitylation, phosphorylation and sumoylation. Acetylation is the most highly studied of these modifications. For example, acetylation of the K14 and K9 lysines of the tail of histone H3 by histone acetyltransferase enzymes (HATs) is generally correlated with transcriptional competence.

One mode of thinking is that this tendency of acetylation to be associated with "active" transcription is biophysical in nature. Because it normally has a positively charged nitrogen at its end, lysine can bind the negatively charged phosphates of the DNA backbone and prevent them from repelling each other. The acetylation event converts the positively charged amine group on the side chain into a neutral amide linkage. This removes the positive charge, causing the DNA to repel itself. When this occurs, complexes like SWI/SNF and other transcriptional factors can bind to the DNA, thus opening it up and exposing it to enzymes like RNA polymerase so transcription of the gene can occur.

In addition, the positively charged tails of histone proteins from one nucleosome may interact with the histone proteins on a neighbouring nucleosome, causing them to pack closely. Lysine acetylation may interfere with these interactions, causing the chromatin structure to open up.

Lysine acetylation may also act as a beacon to recruit other activating chromatin modifying enzymes (and basal transcription machinery as well). Indeed, the bromodomain — a protein segment (domain) that specifically binds acetyl-lysine— is found in many enzymes that help activate transcription, including the SWI/SNF complex (on the protein polybromo). It may be that acetylation acts in this and the previous way to aid in transcriptional activation.

The idea that modifications act as docking modules for related factors is borne out by histone methylation as well. Methylation of lysine 9 of histone H3 has long been associated with constitutively transcriptionally silent chromatin (constitutive heterochromatin). It has been determined that a chromodomain (a domain that specifically binds methyl-lysine) in the transcriptionally repressive protein HP1 recruits HP1 to K9 methylated regions. One example that seems to refute this biophysical model for acetylation is that tri-methylation of histone H3 at lysine 4 is strongly associated with (and required for full) transcriptional activation. Tri-methylation in this case would introduce a fixed positive charge on the tail.

It should be emphasized that differing histone modifications are likely to function in differing ways; acetylation at one position is likely to function differently than acetylation at another position. Also, multiple modifications may occur at the same time, and these modifications may work together to change the behaviour of the nucleosome. The idea that multiple dynamic modifications regulate gene transcription in a systematic and reproducible way is called the histone code.

DNA methylation frequently occurs in repeated sequences, and helps to suppress the expression and mobility of 'transposable elements': Because 5-methylcytosine is chemically very similar to thymidine, CpG sites are frequently mutated and become rare in the genome, except at CpG islands where they remain unmethylated. Epigenetic changes of this type thus have the potential to direct increased frequencies of permanent genetic mutation. DNA methylation patterns are known to be established and modified in response to environmental factors by a complex interplay of at least three independent DNA methyltransferases, DNMT1, DNMT3A and DNMT3B, the loss of any of which is lethal in mice.

DNMT1 is the most abundant methyltransferase in somatic cells, localizes to replication foci, has a 10–40-fold preference for hemimethylated DNA and interacts with the proliferating cell nuclear antigen (PCNA). By preferentially modifying hemimethylated DNA, DNMT1 transfers patterns of methylation to a newly synthesized strand after DNA replication, and therefore is often referred to as the 'maintenance' methyltransferase. DNMT1 is essential for proper embryonic development, imprinting and X-inactivation.

Chromosomal regions can adopt stable and heritable alternative states resulting in bistable gene expression without changes to the DNA sequence. Epigenetic control is often associated with alternative covalent modifications of histones. The stability and heritability of states of larger chromosomal regions are often thought to involve positive feedback where modified nucleosomes recruit enzymes that similarly modify nearby nucleosomes. A simplified stochastic model for this type of epigenetics is found here.

Because DNA methylation and chromatin remodelling play such a central role in many types of epigenic inheritance, the word "epigenetics" is sometimes used as a synonym for these processes. However, this can be misleading. Chromatin remodelling is not always inherited, and not all epigenetic inheritance involves chromatin remodelling.

It has been suggested that the histone code could be mediated by the effect of small RNAs. The recent discovery and characterization of a vast array of small (21- to 26-nt), non-coding RNAs suggests that there is an RNA component, possibly involved in epigenetic gene regulation. Small interfering RNAs can modulate transcriptional gene expression via epigenetic modulation of targeted promoters.

RNA Transcripts and their Encoded Proteins

Sometimes a gene, after being turned on, transcribes a product that (either directly or indirectly) maintains the activity of that gene. For example, Hnf4 and MyoD enhance the transcription of many liver- and muscle-specific genes, respectively, including their own, through the transcription factor activity of the proteins they encode. RNA signalling includes differential recruitment of a hierarchy of generic chromatin modifying complexes and DNA methyltransferases to specific loci by RNAs during differentiation and development.

Other epigenetic changes are mediated by the production of different splice forms of RNA, or by formation of double-stranded RNA (RNAi). Descendants of the cell in which the gene was turned on will inherit this activity, even if the original stimulus for gene-activation is no longer present. These genes are most often turned on or off by signal transduction, although in some systems where syncytia or gap junctions are important, RNA may spread directly to other cells or nuclei by diffusion. A large amount of RNA and protein is contributed to the zygote by the mother during oogenesis or via nurse cells, resulting in maternal effect phenotypes. A smaller quantity of sperm RNA is transmitted from the father, but there is recent evidence that this epigenetic information can lead to visible changes in several generations of offspring.

Prions

Prions are infectious forms of proteins. Proteins generally fold into discrete units which perform distinct cellular functions, but some proteins are also capable of forming an infectious conformational state known as a prion. Although often viewed in the context of infectious disease, prions are more loosely defined by their ability to catalytically convert other native state versions of the same protein to an infectious conformational state. It is in this latter sense that they can be viewed as epigenetic agents capable of inducing a phenotypic change without a modification of the genome.

Fungal prions are considered epigenetic because the infectious phenotype caused by the prion can be inherited without modification of the genome. PSI+ and URE3, discovered in yeast in 1965 and 1971, are the two best studied of this type of prion.

Prions can have a phenotypic effect through the sequestration of protein in aggregates, thereby reducing that protein's activity. In PSI+ cells, the loss of the Sup35 protein (which is involved in termination of translation) causes ribosomes to have a higher rate of read-through of stop codons, an effect which results in suppression of nonsense mutations in other genes. The ability of Sup35 to form prions may be a conserved trait. It could confer an adaptive advantage by giving cells the ability to switch into a PSI+ state and express dormant genetic features normally terminated by premature stop codon mutations.

Structural Inheritance Systems

In ciliates such as *Tetrahymena* and *Paramecium*, genetically identical cells show heritable differences in the patterns of ciliary rows on their cell surface. Experimentally altered patterns can be transmitted to daughter cells. It seems existing structures act as templates for new structures. The mechanisms of such inheritance are unclear, but reasons exist to assume that multicellular organisms also use existing cell structures to assemble new ones.

Functions and Consequences

Development

Somatic epigenetic inheritance, particularly through DNA methylation and chromatin remodelling, is very important in the development of multicellular eukaryotic organisms. The genome sequence is static (with some notable exceptions), but cells differentiate into many different types, which perform different functions, and respond differently to the environment and intercellular signalling. Thus, as individuals develop, morphogens activate or silence genes in an epigenetically heritable fashion, giving cells a "memory". In mammals, most cells terminally differentiate, with only stem cells retaining the ability to differentiate into several cell types ("totipotency" and "multipotency"). In mammals, some stem cells continue producing new differentiated cells throughout life, but mammals are not able to respond to loss of some tissues, for example, the inability to regenerate limbs, which some other animals are capable of. Unlike animals, plant cells do not terminally differentiate, remaining totipotent with the ability to give rise to a new individual plant. While plants do utilise many of the same epigenetic mechanisms as animals, such as chromatin remodelling, it has been hypothesised that plant cells do not have "memories", resetting their gene expression patterns at each cell division using positional information from the environment and surrounding cells to determine their fate.

Medicine

Epigenetics has many and varied potential medical applications. Congenital genetic disease is well understood, and it is also clear that epigenetics can play a role, for example, in the case of Angelman syndrome and Prader-Willi syndrome. These are normal genetic diseases caused by gene deletions or inactivation of the genes, but are unusually common because individuals are essentially hemizygous because of genomic imprinting, and therefore a single gene knock out is sufficient to cause the disease, where most cases would require both copies to be knocked out.

Evolution

Although epigenetics in multicellular organisms is generally thought to be a mechanism involved in differentiation, with epigenetic patterns "reset" when organisms reproduce, there have been some observations of transgenerational epigenetic inheritance (e.g., the phenomenon of paramutation observed in maize). Although most of these multigenerational epigenetic traits are gradually lost over several generations, the possibility remains that multigenerational epigenetics could be another aspect to evolution and adaptation. A sequestered germ line or Weismann barrier is specific to animals, and epigenetic inheritance is expected to be far more common in plants and microbes. These effects may require enhancements to the standard conceptual framework of the modern evolutionary synthesis. Epigenetic features may play a role in short-term adaptation of species by allowing for reversible phenotype variability. The modification of epigenetic features associated with a region of DNA allows organisms, on a multigenerational time scale, to switch between phenotypes that express and repress that particular gene. When the DNA sequence of the region is not mutated, this change is reversible. It has also been speculated that organisms may take advantage of differential mutation rates associated with epigenetic features to control the mutation rates of particular genes.

Epigenetic changes have also been observed to occur in response to environmental exposure—for example, mice given some dietary supplements have epigenetic changes affecting expression of the agouti gene, which affects their fur colour, weight, and propensity to develop cancer.

Transgenerational Epigenetic Observations

A comprehensive review of over 100 cases of transgenerational epigenetic inheritance reported the phenomena in a wide range of organisms including prokaryotes, plants, and animals.

Epigenetic Effects in Humans

Genomic Imprinting and Related Disorders

Some human disorders are associated with genomic imprinting, a phenomenon in mammals where the father and mother contribute different epigenetic patterns for specific genomic loci in their germ cells. The best-known case of imprinting in human disorders is that of Angelman syndrome and Prader-Willi syndrome—both can be produced by the same genetic mutation, chromosome 15q partial deletion, and the particular syndrome that will develop depends on whether the mutation is inherited from the child's mother or from their father. This is due to the presence of genomic imprinting in the region. Beckwith-Wiedemann syndrome is also associated with genomic imprinting, often caused by abnormalities in maternal genomic imprinting of a region on chromosome 11.

Transgenerational Epigenetic Observations

Marcus Pembrey and colleagues also observed in the Φverkalix study that the paternal (but not maternal) grandsons of Swedish boys who were exposed during preadolescence to famine in the 19th century were less likely to die of cardiovascular disease; if food was plentiful then diabetes mortality in the grandchildren increased, suggesting that this was a transgenerational epigenetic inheritance. The opposite effect was observed for females—the paternal (but not maternal) granddaughters of women who experienced famine while in the womb (and their eggs were being formed) lived shorter lives on average.

Cancer and Developmental Abnormalities

A variety of compounds are considered as epigenetic carcinogens— they result in an increased incidence of tumours, but they do not show mutagen activity (toxic compounds or pathogens that cause tumours incident to increased regeneration should also be excluded). Examples include diethylstilbestrol, arsenite, hexachlorobenzene, and nickel compounds.

Many teratogens exert specific effects on the fetus by epigenetic mechanisms. While epigenetic effects may preserve the effect of a teratogen such as diethylstilbestrol throughout the life of an affected child, the possibility of birth defects resulting from exposure of fathers or in second and succeeding generations of offspring has generally been rejected on theoretical grounds and for lack of evidence. However, a range of male-mediated abnormalities have been demonstrated, and more are likely to exist.

FDA label information for Vidaza(tm), a formulation of 5-azacitidine (an unmethylatable analog of cytidine that causes hypomethylation when incorporated into DNA) states that "men should be advised not to father a child" while using the drug, citing evidence in treated male mice of reduced fertility, increased embryo loss, and abnormal embryo development. In rats, endocrine differences were observed in offspring of males exposed to morphine. In mice, second generation effects of diethylstilbesterol have been described occurring by epigenetic mechanisms. In 2008, the National Institutes of Health announced that $190 million had been earmarked for epigenetics research over the next five years. In announcing the funding, government officials noted that epigenetics has the potential to explain mechanisms of aging, human development, and the origins of cancer, heart disease, mental illness, as well as several other conditions. Some investigators, like Randy Jirtle, PhD, of Duke University Medical Centre, think epigenetics may ultimately turn out to have a greater role in disease than genetics.

Twin Studies

Recent studies involving both dizygotic and monozygotic twins have produced some evidence of epigenetic influence in humans.

Epigenetics in Microorganisms

Bacteria make widespread use of postreplicative DNA methylation for the epigenetic control of DNA-protein interactions. Bacteria make use of DNA adenine methylation (rather than DNA cytosine methylation) as an epigenetic signal. DNA adenine methylation is important in bacteria virulence in organisms such as *Escherichia coli, Salmonella, Vibrio, Yersinia, Haemophilus*, and *Brucella*. In *Alphaproteobacteria*, methylation of adenine regulates the cell cycle and couples gene transcription to DNA replication. In *Gammaproteobacteria*, adenine methylation provides signals for DNA replication, chromosome segregation, mismatch repair, packaging of bacteriophage, transposase activity and regulation of gene expression.

The filamentous fungus *Neurospora crassa* is a prominent model system for understanding the control and function of cytosine methylation. In this organisms, DNA methylation is associated with relics of a genome defence system called RIP (repeat-induced point mutation) and silences gene expression by inhibiting transcription elongation.

The yeast prion PSI is generated by a conformational change of a translation termination factor, which is then inherited by daughter cells. This can provide a survival advantage under adverse conditions. This is an example of epigenetic regulation enabling unicellular organisms to respond rapidly to environmental stress. Prions can be viewed as epigenetic agents capable of inducing a phenotypic change without modification of the genome.

DNA Methyltransferases

In mammalian cells, DNA methylation occurs mainly at the C5 position of CpG dinucleotides and is carried out by two general classes of enzymatic activities – maintenance methylation and *de novo* methylation.

Maintenance methylation activity is necessary to preserve DNA methylation after every cellular DNA replication cycle. Without the DNA methyltransferase (DNMT), the replication machinery itself would produce daughter strands that are unmethylated and, over time, would lead to passive demethylation. DNMT1 is the proposed maintenance methyltransferase that is responsible for copying DNA methylation patterns to the daughter strands during DNA replication. Mouse models with both copies of DNMT1 deleted are embryonic lethal at approximately day 9, due to the requirement of DNMT1 activity for development in mammalian cells.

It is thought that DNMT3a and DNMT3b are the *de novo* methyltransferases that set up DNA methylation patterns early in development. DNMT3L is a protein that is homologous to the other DNMT3s but has no catalytic activity. Instead, DNMT3L assists the *de novo* methyltransferases by increasing their ability to bind to DNA and stimulating their activity. Finally, DNMT2 (TRDMT1) has been identified as a DNA methyltransferase homolog, containing all 10 sequence motifs common to all DNA methyltransferases; however, DNMT2 (TRDMT1) does not methylate DNA but instead methylates cytosine-38 in the anticodon loop of aspartic acid transfer RNA.

Since many tumour suppressor genes are silenced by DNA methylation during carcinogenesis, there have been attempts to re-express these genes by inhibiting the DNMTs. 5-Aza-2'-deoxycytidine (decitabine) is a nucleoside analog that inhibits DNMTs by trapping them in a covalent complex on DNA by preventing the β-elimination step of catalysis, thus resulting in the enzymes' degradation. However, for decitabine to be active, it must be incorporated into the genome of the cell, which can cause mutations in the daughter cells if the cell does not die. In addition, decitabine is toxic to the bone marrow, which limits the size of its therapeutic window. These pitfalls have led to the development of antisense RNA therapies that target the DNMTs by degrading their mRNAs and preventing their translation. However, it is currently unclear whether targeting DNMT1 alone is sufficient to reactivate tumour suppressor genes silenced by DNA methylation.

In Plants

Significant progress has been made in understanding DNA methylation in the model plant *Arabidopsis thaliana*. DNA methylation in plants differs from that of mammals: while DNA methylation in mammals mainly occurs on the cytosine nucleotide in a CpG site, in plants the cytosine can be methylated at CpG, CpNpG, and CpNpN sites, where N represents any nucleotide but guanine.

The principal *Arabidopsis* DNA methyltransferase enzymes, which transfer and covalently attach methyl groups onto DNA, are DRM2, MET1, and CMT3. Both the DRM2 and MET1 proteins share significant homology to the mammalian methyltransferases DNMT3 and DNMT1, respectively, whereas the CMT3 protein is unique to the plant kingdom. There are currently two classes of DNA methyltransferases: 1) the *de novo* class, or enzymes that create new methylation marks on the DNA, and 2) a maintenance class that recognizes the methylation marks on the parental strand of DNA and transfers new methylation to the daughters strands after DNA replication.

DRM2 is the only enzyme that has been implicated as a *de novo* DNA methyltransferase. DRM2 has also been shown, along with MET1 and CMT3 to be involved in maintaining methylation marks through DNA replication. Other DNA methyltransferases are expressed in plants but have no known function.

It is not clear how the cell determines the locations of *de novo* DNA methylation, but evidence suggests that, for many (though not all) locations, RNA-directed DNA methylation (RdDM) is involved. In RdDM, specific RNA transcripts are produced from a genomic DNA template, and this RNA forms secondary structures called double-stranded RNA molecules.

The double-stranded RNAs, through either the small interfering RNA (siRNA) or microRNA (miRNA) pathways direct de-novo DNA methylation of the original genomic location that produced the RNA. This sort of mechanism is thought to be important in cellular defence against RNA viruses and/or transposons, both of which often form a double-stranded RNA that can be mutagenic to the host genome. By methylating their genomic locations, through an as yet poorly-understood mechanism, they are shut off and are no longer active in the cell, protecting the genome from their mutagenic effect.

In Fungi

It can be seen that many fungi have low levels (0.1 to 0.5%) of cytosine methylation, whereas other fungi have as much as 5% of the genome methylated.

This value seems to vary both among species and among isolates of the same species. There is also evidence that DNA methylation may be involved in state-specific control of gene expression in fungi. Although brewers' yeast (*Saccharomyces*) and fission yeast (*Schizosaccharomyces*) have very little DNA methylation, the model filamentous fungus *Neurospora crassa* has a well-characterized methylation system. Several genes control methylation in *Neurospora* and mutation of the DNA methyl transferase, *dim-2*, eliminates all DNA methylation but does not affect growth or sexual reproduction. While the *Neurospora* genome has very little repeated DNA, half of the methylation occurs in repeated DNA including transposon relics and centromeric DNA. The ability to evaluate other important phenomena in a DNA methylase-deficient genetic background makes *Neurospora* an important system in which to study DNA methylation.

In Bacteria

Adenine or cytosine methylation is part of the restriction modification system of many bacteria, in which specific DNA sequences are methylated periodically throughout the genome. A methylase is the enzyme that recognizes a specific sequence and methylates one of the bases in or near that sequence. Foreign DNAs (which are not methylated in this manner) that are introduced into the cell are degraded by sequence-specific restriction enzymes and cleaved. Bacterial genomic DNA is not recognized by these restriction enzymes. The methylation of native DNA acts as a sort of primitive immune system, allowing the bacteria to protect themselves from infection by bacteriophage.

E. coli DNA adenine methyltransferase (Dam) is an enzyme of ~32 kDa that does not belong to a restriction/modification system. The target recognition sequence for *E. coli* Dam is GATC, as the methylation occurs at the N6 position of the adenine in this sequence (G meATC). The three base pairs flanking each side of this site also influence DNA–Dam binding. Dam plays several key roles in bacterial processes, including mismatch repair, the timing of DNA replication, and gene expression. As a result of DNA replication, the status of GATC sites in the *E. coli* genome changes from fully methylated to hemimethylated. This is because adenine introduced into the new DNA strand is unmethylated. Re-methylation occurs within two to four seconds, during which time replication errors in the new strand are repaired. Methylation, or its absence, is the marker that allows the repair apparatus of the cell to differentiate between the template and nascent strands. It has been shown that altering Dam activity in bacteria results in increased spontaneous mutation rate. Bacterial viability is compromised in dam mutants that also lack certain other DNA repair enzymes, providing further evidence for the role of Dam in DNA repair. One region of the DNA that keeps its hemimethylated status for longer is the origin of replication, which has an abundance of GATC sites. This is central to the bacterial mechanism for timing DNA replication. SeqA binds to the origin of replication, sequestering it and thus preventing methylation. Because hemimethylated origins of replication are inactive, this mechanism limits DNA replication to once per cell cycle.

Expression of certain genes, for example those coding for pilus expression in *E. coli*, is regulated by the methylation of GATC sites in the promoter region of the gene operon. The cells' environmental conditions just after DNA replication determine whether Dam is blocked from methylating a region proximal to or distal from the promoter region. Once the pattern of methylation has been created, the pilus gene transcription is locked in the on or off position until the DNA is again replicated. In *E. coli*, these pilus operons have important roles in virulence in urinary tract infections. It has been proposed that inhibitors of Dam may function as antibiotics. On the other hand DNA cytosine methylase targets CCAGG and CCTGG sites to methylate cytosine at the C5 position (C meC(A/T)GG). The other methylase enzyme, EcoKI, causes mehtylation of adenine in the sequences AAC(N6A)GTGC and GCAC(N6A)GTT.

Most strains used by molecular biologists are derivatives of K-12, and possess both Dam and Dcm, but there are commercially available strains which possess dam-/dcm- activity. In fact it is possible to unmethylate the DNA extracted from dam+/dcm+ strains by transforming into dam-/dcm- strains. This would help digest sequences that are not being recognized by methylation-sensitive restriction enzymes.

Detection

DNA methylation can be detected by the following assays currently used in scientific research:

- Methylation-Specific PCR (MSP), which is based on a chemical reaction of sodium bisulfite with DNA that converts unmethylated cytosines of CpG dinucleotides to uracil or UpG, followed by traditional PCR. However, methylated cytosines will not be converted in this process, and primers are designed to overlap the CpG site of interest, which allows one to determine methylation status as methylated or unmethylated.

- The HELP assay, which is based on restriction enzymes' differential ability to recognize and cleave methylated and unmethylated CpG DNA sites.

- ChIP-on-chip assays, which is based on the ability of commercially prepared antibodies to bind to DNA methylation-associated proteins like MCP2.

- Restriction landmark genomic scanning, a complicated and now rarely-used assay based upon restriction enzymes' differential recognition of methylated and unmethylated CpG sites; the assay is similar in concept to the HELP assay.

- Methylated DNA immunoprecipitation (MeDIP), analogous to chromatin immunoprecipitation, immunoprecipitation is used to isolate methylated DNA fragments for input into DNA detection methods such as DNA microarrays (MeDIP-chip) or DNA sequencing (MeDIP-seq).

- Molecular break light assay for DNA adenine methyltransferase activity – an assay that relies on the specificity of the restriction enzyme DpnI for fully methylated (adenine methylation) GATC sites in an oligonucleotide labelled with a fluorophore and quencher. The adenine methyltransferase methylates the oligonucleotide making it a substrate for DpnI. Cutting of the oligonucleotide by DpnI gives rise to a fluorescence increase.

Histone

Histones are highly alkaline proteins found in eukaryotic cell nuclei, which package and order the DNA into structural units called nucleosomes. They are the chief protein components of chromatin, act as spools around which DNA winds, and play a role in gene regulation. Without histones, the unwound DNA in chromosomes would be very long (a length to width ratio of more than 10 million to one in human DNA). For example, each human cell has about 1.8 meters of DNA, but wound on the histones it has about 90 millimeters of chromatin, which, when duplicated and condensed during mitosis, result in about 120 micrometers of chromosomes.

Classes

Histones "are highly conserved and can be grouped into five major classes: H1/H5, H2A, H2B, H3, and H4". These are organised into two super-classes as follows:

- core histones – H2A, H2B, H3 and H4
- linker histones – H1 and H5.

Two of each of the core histones assemble to form one octameric nucleosome core particle by wrapping 147 base pairs of DNA around the protein spool in 1.65 left-handed super-helical turn. The linker histone H1 binds the nucleosome and the entry and exit sites of the DNA, thus locking the DNA into place and allowing the formation of higher order structure. The most basic such formation is the 10 nm fiber or beads on a string conformation.

The following is a list of human histone proteins:

Super family	Family	Subfamily	Members
Linker	H1	H1F	H1F0, H1FNT, H1FOO, H1FX
		H1H1	HIST1H1A, HIST1H1B, HIST1H1C, HIST1H1D, HIST1H1E, HIST1H1T
Core	H2A	H2AF	H2AFB1, H2AFB2, H2AFB3, H2AFJ, H2AFV, H2AFX, H2AFY, H2AFY2, H2AFZ
		H2A1	HIST1H2AA, HIST1H2AB, HIST1H2AC, HIST1H2AD, HIST1H2AE, HIST1H2AG, HIST1H2AI, HIST1H2AJ, HIST1H2AK, HIST1H2AL, HIST1H2AM
		H2A2	HIST2H2AA3, HIST2H2AC
	H2B	H2BF	H2BFM, H2BFO, H2BFS, H2BFWT
		H2B1	HIST1H2BA, HIST1H2BB, HIST1H2BC, HIST1H2BD, HIST1H2BE, HIST1H2BF, HIST1H2BG, HIST1H2BH, HIST1H2BI, HIST1H2BJ, HIST1H2BK, HIST1H2BL, HIST1H2BM, HIST1H2BN, HIST1H2BO
		H2B2	HIST2H2BE
	H3	H3A1	HIST1H3A, HIST1H3B, HIST1H3C, HIST1H3D, HIST1H3E, HIST1H3F, HIST1H3G, HIST1H3H, HIST1H3I, HIST1H3J
		H3A2	HIST2H3C
		H3A3	HIST3H3
	H4	H41	HIST1H4A, HIST1H4B, HIST1H4C, HIST1H4D, HIST1H4E, HIST1H4F, HIST1H4G, HIST1H4H, HIST1H4I, HIST1H4J, HIST1H4K, HIST1H4L
		H44	HIST4H4

This involves the wrapping of DNA around nucleosomes with approximately 50 base pairs of DNA separating each pair of nucleosomes (also referred to as linker DNA). The assembled histones and DNA is called chromatin. Higher-order structures include the 30 nm fiber (forming an irregular zigzag) and 100 nm fiber, these being the structures found in normal cells. During mitosis and meiosis, the condensed chromosomes are assembled through interactions between nucleosomes and other regulatory proteins.

Structure

The nucleosome core is formed of two H2A-H2B dimers and a H3-H4 tetramer, forming two nearly symmetrical halves by tertiary structure (C2 symmetry; one macromolecule is the mirror image of the other). The H2A-H2B dimers and H3-H4 tetramer also show pseudodyad symmetry. The 4 'core' histones (H2A, H2B, H3 and H4) are relatively similar in structure and are highly conserved through evolution, all featuring a 'helix turn helix turn helix' motif (which allows the easy dimerisation). They also share the feature of long 'tails' on one end of the amino acid structure - this being the location of post-translational modification. Using an electron paramagnetic resonance spin-labelling technique, British researchers measured the distances between the spools around which eukaryotic cells wind their DNA. They determined the spacings range from 59 to 70 Å.

In all, histones make five types of interactions with DNA:

- Helix-dipoles from alpha-helices in H2B, H3, and H4 cause a net positive charge to accumulate at the point of interaction with negatively charged phosphate groups on DNA
- Hydrogen bonds between the DNA backbone and the amide group on the main chain of histone proteins
- Nonpolar interactions between the histone and deoxyribose sugars on DNA
- Salt bridges and hydrogen bonds between side chains of basic amino acids (especially lysine and arginine) and phosphate oxygens on DNA
- Non-specific minor groove insertions of the H3 and H2B N-terminal tails into two minor grooves each on the DNA molecule.

The highly basic nature of histones, aside from facilitating DNA-histone interactions, contributes to the water solubility of histones. Histones are subject to post translational modification by enzymes primarily on their N-terminal tails, but also in their globular domains Such modifications include methylation, citrullination, acetylation, phosphorylation, SUMOylation, ubiquitination, and ADP-ribosylation. This affects their function of gene regulation. In general, genes that are active have less bound histone, while inactive genes are highly associated with histones during interphase It also appears that the structure of histones has been evolutionarily conserved, as any deleterious mutations would be severely maladaptive.

Function

Compacting DNA Strands

Histones act as spools around which DNA winds. This enables the compaction necessary to fit the large genomes of eukaryotes inside cell nuclei: the compacted molecule is 40,000 times shorter than an unpacked molecule.

Chromatin Regulation

Histones undergo posttranslational modifications which alter their interaction with DNA and nuclear proteins. The H3 and H4 histones have long tails protruding from the nucleosome which can be covalently modified at several places. Modifications of the tail include methylation, acetylation, phosphorylation, ubiquitination, SUMOylation, citrullination and ADP-ribosylation. The core of the histones H2A, H2B and H3 can also be modified. Combinations of modifications are thought to constitute a code, the so-called "histone code". Histone modifications act in diverse biological processes such as gene regulation, DNA repair and chromosome condensation (mitosis).

The common nomenclature of histone modifications is:

- The name of the histone (*e.g.* H3)
- The single letter amino acid abbreviation (*e.g.* K for Lysine) and the amino acid position in the protein
- The type of modification (Me: methyl, P: phosphate, Ac: acetyl, Ub: ubiquitin).

So H3K4me1 denotes the monomethylation of the 4th residue (a lysine) from the start (i.e., the N-terminal) of the H3 protein.

History

Histones were discovered in 1884 by Albrecht Kossel. The word "histone" dates from the late 19th century and is from the German

"Histon", of uncertain origin: perhaps from Greek *histanai* or from *histos*. Until the early 1990s, histones were dismissed as merely packing material for nuclear DNA. During the early 1990s, the regulatory functions of histones were discovered.

The discovery of the H5 histone appears to date back to 1970's, and in classification it has been grouped with H1.

Conservation Across Species

Histones are found in the nuclei of eukaryotic cells, and in certain Archaea, namely Euryarchaea, but not in bacteria. Archaeal histones may well resemble the evolutionary precursors to eukaryotic histones. Histone proteins are among the most highly conserved proteins in eukaryotes, emphasizing their important role in the biology of the nucleus. In contrast mature sperm cells largely use protamines to package their genomic DNA, most likely to achieve an even higher packaging ratio.

Core histones are highly conserved proteins, that is, there are very few differences among the amino acid sequences of the histone proteins of different species. Linker histone usually has more than one form within a species and is also less conserved than the core histones.

There are some *variant* forms in some of the major classes. They share amino acid sequence homology and core structural similarity to a specific class of major histones but also have their own feature that is distinct from the major histones. These *minor histones* usually carry out specific functions of the chromatin metabolism. For example, histone H3-like CenpA is a histone only associated with the centromere region of the chromosome. Histone H2A variant H2A.Z is associated with the promoters of actively transcribed genes and also involved in the prevention of the spread of silent heterochromatin. Another H2A variant H2A.X binds to the DNA with double strand breaks and marks the region undergoing DNA repair. Histone H3.3 is associated with the body of actively transcribed genes.

Histone Code

The Histone Code is hypothesized to be a code consisting of covalent histone tail modifications. Together with other modifications such as DNA methylation it is part of the epigenetic code·

The main role of histones is in associating with DNA to form nucleosomes, which themselves bundle to form chromatin fibers. They are globular proteins with a flexible N-terminus (taken to be the tail) that protrudes from the nucleosome. The tail modifications play an important role in the chromatin structure, while the chromatin structure plays an important role in regulation of gene expression

The Hypothesis

The hypothesis is that chromatin-DNA interactions are guided by combinations of histone modifications. While it is accepted that modifications (such as methylation, acetylation, ADP-ribosylation, ubiquitination and phosphorylation) to histone tails alter chromatin structure, a complete understanding of the precise mechanisms by which these alterations to histone tails influence DNA-histone interactions remains elusive.

However, some specific examples have been worked out in detail. For example, phosphorylation of serine residues 10 and 28 on histone H3 is a marker for chromosomal condensation. Similarly, the combination of phosphorylation of serine residue 10 and acetylation of a lysine residue 14 on histone H3 is a tell-tale sign of active transcription.

Modifications

Possible modifications to the tails are:

- Acetylation - by HAT (Histone Acetyl Transferase); deacetylation - by HDAC (Histone Deacetylase)

Deacetylation allows tight arrangement of chromatin, preventing gene expression, while acetylation may occur to open up the chromatin.

- Methylation.

Methylation of lysines H3K4 and H3K36 is correlated with transcriptional activation while demethylation of H3K4 is correlated with silencing of the genomic region. Methylation of lysines H3K9 and H3K27 is correlated with transcriptional repression Particularly, H3K9me3 is highly correlated with constitutive heterochromatin.

- Phosphorylation
- Ubiquitination
- ADP-ribosylation.

Histone-Modifying Enzymes

The packaging of the eukaryotic genome into highly condensed chromatin makes it inaccessible to the factors required for gene transcription, DNA replication, recombination and repair. Eukaryotes have developed intricate mechanisms to overcome this repressive barrier imposed by the chromatin. The nucleosome is composed of an octamer of the four core histones (H3, H4, H2A, H2B) around which 147 base pairs of DNA are wrapped. Several distinct classes of enzyme can modify histones at multiple sites. The figure on the right enlists those histone-modifying enzymes whose specificity has been determined. There are at least eight distinct types of modifications found on histones. Enzymes have been identified for acetylation methylation demethylation phosphorylation ubiquitination sumoylation ADP-ribosylation deimination and proline isomerization

Histone Deacetylase

Histone deacetylases (HDAC) (EC number 3.5.1) are a class of enzymes that remove acetyl groups from an ε-N-acetyl lysine amino acid on a histone. Its action is opposite to that of histone acetyltransferase. HDAC proteins are now also being referred to as lysine deacetylases (KDAC), as to more precisely describe their function rather than their target, which also includes numerous non-histone proteins.

Subtypes

HDAC proteins occur in four groups based on function and DNA sequence similarity. The first two groups are considered "classical" HDACs whose activities are inhibited by trichostatin A (TSA), whereas the third group is a family of NAD+-dependent proteins not affected by TSA.

Homologues to these three groups are found in yeast having the names: reduced potassium dependency 3 (Rpd3), which corresponds to Class I; histone deacetylase 1 (hda1), corresponding to Class II; and silent information regulator 2 (Sir2); corresponding to Class III. The fourth group is considered an atypical category of its own, based solely on DNA sequence similarity to the others.

Subcellular Distribution

Within the Class I HDACs, HDAC 1, 2, and 8 are found primarily in the nucleus, whereas HDAC 3 is found in both the nucleus and the cytoplasm, and is also membrane-associated. Class II HDACs (HDAC 4, 5, 6, 7 9, and 10) are able to shuttle in and out of the nucleus, depending on different signals.

HDAC 6 is a cytoplasmic, microtuble-associated enzyme. HDAC 6 deacetylates tubulin, Hsp90, and cortactin, and forms complexes with other partner proteins, and is, therefore, involved in a variety of biological processes.

Function

Histone tails are normally positively charged due to amine groups present on their lysine and arginine amino acids. These positive charges help the histone tails to interact with and bind to the negatively charged phosphate groups on the DNA backbone. Acetylation, which occurs normally in a cell, neutralizes the positive charges on the histone by changing amines into amides and decreases the ability of the histones to bind to DNA.

This decreased binding allows chromatin expansion, permitting genetic transcription to take place. Histone deacetylases remove those acetyl groups, increasing the positive charge of histone tails and encouraging high-affinity binding between the histones and DNA backbone. The increased DNA binding condenses DNA structure, preventing transcription.

Histone deacetylase is involved in a series of pathways within the living system. According to the *Kyoto Encyclopedia of Genes and Genomes* (KEGG), these are:

- Environmental information processing; signal transduction; notch signalling pathway PATH:ko04330
- Cellular processes; cell growth and death; cell cycle PATH:ko04110
- Human diseases; cancers; chronic myeloid leukemia PATH:ko05220.

Histone acetylation plays an important role in the regulation of gene expression. Hyperacetylated chromatin is transcriptionally active, and hypoacetylated chromatin is silent. A study on mice found that a specific subset of mouse genes (7%) was deregulated in the absence of HDAC1. Their study also found a regulatory crosstalk between HDAC1 and HDAC2 and suggest a novel function for HDAC1 as a transcriptional coactivator. HDAC1 expression was found to be increased in the prefrontal cortex of schizophrenia subjects, negatively correlating with the expression of GAD67 mRNA.

It is a mistake to regard HDACs solely in the context of regulating gene transcription by modifying histones and chromatin structure, although that appears to be the predominant function. The function, activity, and stability of proteins can be controlled by post-translational modifications. Protein phosphorylation is perhaps the most widely studied and understood modification in which certain amino acid residues are phosphorylated by the action of protein kinases or dephosphorylated by the action of phosphatases. The acetylation of lysine residues is emerging as an analogus mechanism, in which non-histone proteins are acted on by acetylases and deacetylases. It is in this context that HDACs are being found to interact with a variety of non-histone proteins—some of these are transcription factors and co-regulators, some are not. Note the following four examples:

- HDAC6 is associated with aggresomes. Misfolded protein aggregates are tagged by ubiquitination and removed from the cytoplasm by dynein motors via the microtubule network to an organelle termed the aggresome. HDAC 6 binds polyubiquitinated misfolded proteins and links to dynein motors, thereby allowing the misfolded protein cargo to be physically transported to chaperones and proteasomes for subsequent destruction.

- PTEN is an important phosphatase involved in cell signalling via phosphoinositols and the AKT/PI3 kinase pathway. PTEN is subject to complex regulatory control via phosphorylation, ubiquitination, oxidation and acetylation. Acetylation of PTEN by the histone acetyltransferase p300/CBP-associated factor (PCAF) can stimulate its activity; on the converse, deacetylation of PTEN by SIRT1 deacetylase and, by HDAC1, can repress its activity.

- APE1/Ref-1 (APEX1) is a multifunctional protein possessing both DNA repair activity (on abasic and single-strand break sites) and transcriptional regulatory activity associated with oxidative stress. APE1/Ref-1 is acetylated by PCAF; on the converse, it is stably associated with and deacetylated by Class I HDACs. The acetylation state of APE1/Ref-1 does not appear to affect its DNA repair activity, but it does regulate its transcriptional activity such as its ability to bind to the PTH promoter and initiate transcription of the parathyroid hormone gene.

- NF-κB is a key transcription factor and effector molecule involved in responses to cell stress, consisting of a p50/p65 heterodimer. The p65 subunit is controlled by acetylation via PCAF and by deacetylation via HDAC3 and HDAC6.

These are just some examples of constantly emerging non-histone, non-chromatin roles for HDACs.

HDAC Inhibitors

Histone deacetylase inhibitors (HDIs) have a long history of use in psychiatry and neurology as mood stabilizers and anti-epilectics, for example, valproic acid. In more recent times, HDIs are being studied as a mitigator or treatment for neurodegenerative diseases. Also in recent years, there has been an effort to develop HDIs for cancer therapy, and Vorinostat (SAHA) has recently been approved for treatment of cutaneous T cell lymphoma (CTCL). The exact mechanisms by which the compounds may work are unclear, but epigenetic pathways are proposed.

In addition, a clinical trial is studying valproic acid effects on the latent pools of HIV in infected persons. HDIs are currently being investigated as chemosensitizers for cytotoxic chemotherapy or radiation therapy, or in association with DNA methylation inhibitors based on in vitro synergy. Batty N; Malouf, GG; Issa, JP (August 2009). "Histone deacetylase inhibitors as anti-neoplastic agents". *Cancer Letters* 280 (2): 190–200. doi:10.1016/j.canlet.2009.03.013. PMID 19345475.

HDAC inhibitors have effects on non-histone proteins that are related to acetylation. HDIs can alter the degree of acetylation of these molecules and, therefore, increase or repress their activity. For the four examples given above on HDACs acting on non-histone proteins, in each of those instances the HDAC inhibitor Trichostatin A (TSA) blocks the effect. HDIs have been shown to alter the activity of many transcription factors, including ACTR, cMyb, E2F1, EKLF, FEN 1, GATA, HNF-4, HSP90, Ku70, NFκB, PCNA, p53, RB, Runx, SF1 Sp3, STAT, TFIIE, TCF, YY1.

Family

Together with the acetylpolyamine amidohydrolases and the acetoin utilization proteins, the histone deacetylases form an ancient protein superfamily known as the histone deacetylase superfamily. Histone deacetylases, acetoin utilization proteins and acetylpolyamine amidohydrolases are members of an ancient protein superfamily.

Classes of HDACs in Higher Eukaryotes

HDACs, are classified in four classes depending on sequence identity and domain organization:

- Class I
  - — HDAC1, HDAC2, HDAC3, HDAC8
- Class II
  - — HDAC4, HDAC5, HDAC6, HDAC7A, HDAC9, HDAC10
- Class III
  - — Homologs of Sir2 in the yeast *Saccharomyces cerevisiae*
  - — sirtuins in mammals (SIRT1, SIRT2, SIRT3, SIRT4, SIRT5, SIRT6, SIRT7)
- Class IV
  - — HDAC11.

Histone Deacetylase Inhibitor

Histone deacetylase inhibitors (HDAC inhibitors, HDI) are a class of compounds that interfere with the function of histone deacetylase.

Cellular Biochemistry/pharmacology

To carry out gene expression, a cell must control the coiling and uncoiling of DNA around histones. This is accomplished with the assistance of histone acetylases (HAT), which acetylate the lysine residues in core histones leading to a less compact and more transcriptionally active chromatin, and, on the converse, the actions of histone deacetylases (HDAC), which remove the acetyl groups from the lysine residues leading to the formation of a condensed and transcriptionally silenced chromatin. Reversible modification of the terminal tails of core histones constitutes the major epigenetic mechanism for remodelling higher-order chromatin structure and controlling gene expression. HDAC inhibitors (HDI) block this action and can result in hyperacetylation of histones, thereby affecting gene expression.

"The histone deacetylase inhibitors are a new class of cytostatic agents that inhibit the proliferation of tumour cells in culture and in vivo by inducing cell cycle arrest, differentiation and/or apoptosis. Histone acetylation and deacetylation play important roles in the modulation of chromatin topology and the regulation of gene transcription. Histone deacetylase inhibition induces the accumulation of hyperacetyl-ated nucleosome core histones in most regions of chromatin but affects the expression of only a small subset of genes, leading to transcriptional activation of some genes, but repression of an equal or larger number of other genes. Non-histone proteins such as transcription factors are also targets for acetylation with varying functional effects. Ace-tylation enhances the activity of some transcription factors such as the tumour suppressor p53 and the erythroid differentiation factor GATA-1 but may repress transcriptional activity of others including T cell factor and the co-activator ACTR. Recent studies [...] have shown that the estrogen receptor alpha (ERalpha) can be hyperacetylated in response to histone deacetylase inhibition, suppressing ligand sensitivity and regulating transcriptional activation by histone deacetylase inhibitors. Conservation of the acetylated ERalpha motif in other nuclear receptors suggests that acetylation may play an important regulatory role in diverse nuclear receptor signalling functions. A number of structurally diverse histone deacetylase inhibitors have shown potent antitumor efficacy with little toxicity in vivo in animal models. Several compounds are currently in early phase clinical development as potential treatments for solid and hematological cancers both as monotherapy and in combination with cytotoxics and differentiation agents."

HDAC Classification

HDACs are classified in four groups based on their homology to yeast histone deacetylases:

- Class I, which includes HDAC1, -2, -3 and -8 are related to yeast RPD3 gene;
- Class II, which includes HDAC4, -5, -6, -7, -9 and -10 are related to yeast Hda1 gene;
- Class III, also known as the sirtuins are related to the Sir2 gene and include SIRT1-7, and
- Class IV, which contains only HDAC11 has features of both Class I and II.

HDI Classification

The "classical" HDIs act exclusively on Class I and Class II HDACs by binding to the zinc-containing catalytic domain of the HDACs. These classical HDIs fall into several groupings, in order of decreasing potency

1. hydroxamic acids (or hydroxamates), such as trichostatin A,
2. cyclic tetrapeptides (such as trapoxin B), and the depsipeptides,
3. benzamides,
4. electrophilic ketones, and

5. the aliphatic acid compounds such as phenylbutyrate and valproic acid.

"Second-generation" HDIs include the hydroxamic acids vorinostat (SAHA), belinostat (PXD101), LAQ824, and panobinostat (LBH589); and the benzamides : entinostat (MS-275), CI994, and mocetinostat (MGCD0103). The sirtuin Class III HDACs are dependent on NAD+ and are, therefore, inhibited by nicotinamide, as well derivatives of NAD, dihydrocoumarin, naphthopyranone, and 2-hydroxynaphaldehydes.

Additional Functions

HDIs should not be considered to act solely as enzyme inhibitors of HDACs. A large variety of nonhistone transcription factors and transcriptional co-regulators are known to be modified by acetylation. HDIs can alter the degree of acetylation nonhistone effector molecules and, therefore, increase or repress the transcription of genes by this mechanism. Examples include: ACTR, cMyb, E2F1, EKLF, FEN 1, GATA, HNF-4, HSP90, Ku70, NF-κB, PCNA, p53, RB, Runx, SF1 Sp3, STAT, TFIIE, TCF, YY1, etc.

Uses

Psychiatry and Neurology

HDIs have a long history of use in psychiatry and neurology as mood stabilzers and anti-epileptics. The prime example of this is valproic acid, marketed as a drug under the trade names *Depakene*, *Depakote*, and *Divalproex*. In more recent times, HDIs are being studied as a mitigator for neurodegenerative diseases such as Alzheimer's disease and Huntington's disease. Enhancement of memory formation is increased in mice given the HDIs sodium butyrate or SAHA, or by genetic knockout of the HDAC2 gene in mice. While that may have relevance to Alzheimer's disease, it was shown that some cognitive deficits were restored in actual transgenic mice that have a model of Alzheimer's disease (3xTg-AD) by orally administered nicotinamide, a competitive HDI of Class III sirtuins.

Cancer Treatment

Also in recent years, there has been an effort to develop HDIs as a cancer treatment or adjunct The exact mechanisms by which the compounds may work are unclear, but epigenetic pathways are proposed. Richon et al. found that HDAC inhibitors can induce p21

(WAF1) expression, a regulator of p53's tumour suppressor activity. HDACs are involved in the pathway by which the retinoblastoma protein (pRb) suppresses cell proliferation. The pRb protein is part of a complex that attracts HDACs to the chromatin so that it will deacetylate histones. HDAC1 negatively regulates the cardiovascular transcription factor Kruppel-like factor 5 through direct interaction. Estrogen is well-established as a mitogenic factor implicated in the tumorigenesis and progression of breast cancer via its binding to the estrogen receptor alpha (ERα). Recent data indicate that chromatin inactivation mediated by HDAC and DNA methylation is a critical component of ERα silencing in human breast cancer cells.

Approved
- Vorinostat was licenced by the U.S. FDA in October 2006 for the treatment of cutaneous T cell lymphoma (CTCL).
- Romidepsin (trade name Istodax) was licenced by the US FDA in Nov 2009 for cutaneous T-cell lymphoma (CTCL),

Started Phase III Clinical Trials
- Panobinostat (LBH589) is in clinical trials for various cancers including a phase III trial for cutaneous T cell lymphoma (CTCL).

Started Phase II Clinical Trials
- Valproic acid is under investigation for various cancers including leukemia.
- Belinostat (PXD101) has had a phase II trial for relapsed ovarian cancer.
- PCI-24781 has started a phase II trial for sarcoma and another for lymphoma
- SB939 starting a phase II trial for Recurrent or Metastatic Prostate Cancer (HRPC).
- Entinostat (MS-275) in phase II for Hodgkin lymphoma, lung cancer and breast cancer.

Started Phase I Clinical Trials
- Mocetinostat (MGCD0103) is undergoing clinical trials for treatment of various cancers (including follicular lymphoma, Hodgkin lymphoma and acute myeloid leukemia).
- AR-42 has started clinical trials in 2010 for various cancers (relapsed or treatment-resistant multiple myeloma, chronic lymphocytic leukemia or lymphoma).

- CUDC-101 intended for cancer, has started clinical trials it also inhibits EGFR and HER2. Early results reported in Nov 2010.
- Also sulforaphane, FK228, ITF2357.

Inflammatory Diseases

Trichostatin A (TSA) and others are being investigated as anti-inflammatory agents.

Other Diseases

- ITF2357 is under investigation for treatment of polycythemia vera (PV), essential thrombocythemia (ET) and myelofibrosis (MF).

Histone Methyltransferase

Histone methyltransferases (HMT) are enzymes, histone-lysine N-methyltransferase and histone-arginine N-methyltransferase, that catalyse the transfer of one to three methyl groups from the cofactor S-Adenosyl methionine to lysine and arginine residues of histone proteins. *These proteins* often contain a SET (Su(var)3-9, Enhancer of Zeste, Trithorax) domain, however the recently discovered HMT Dot1 lacks the characteristic SET domain.

Role in Gene Regulation

Histone methylation serves in epigenetic gene regulation. Methylated histones bind DNA more tightly, which inhibits transcription. Methylated histones can either repress or activate transcription as different experimental findings suggest.

RNA Polymerases

A polymerase (EC 2.7.7.6/7/19/48/49) is an enzyme whose central function is associated with polymers of nucleic acids such as RNA and DNA. The primary function of a polymerase is the polymerization of new DNA or RNA against an existing DNA or RNA template in the processes of replication and transcription. In association with a cluster of other enzymes and proteins, they take nucleotides from solvent, and catalyse the synthesis of a polynucleotide sequence against a nucleotide template strand using base-pairing interactions. It is an accident of history that the enzymes responsible for the catalytic production of other biopolymers are not also referred to as polymerases.

One particular polymerase, from the thermophilic bacterium, *Thermus aquaticus* (*Taq*) (PDB 1BGX, EC 2.7.7.7) is of vital commercial importance due to its use in the polymerase chain reaction, a widely used technique of molecular biology.

Other well-known polymerases include:

- Terminal Deoxynucleotidyl Transferase (TDT), which lends diversity to antibody heavy chains
- Reverse Transcriptase, an enzyme used by RNA retroviruses like HIV, which is used to create a complementary strand to the preexisting strand of viral RNA before it can be integrated into the DNA of the host cell. It is also a major target for antiviral drugs.

RNA Polymerase I

RNA polymerase I (also called Pol I) is, in eukaryotes, the only enzyme that transcribes ribosomal RNA (excluding 5S rRNA, which is synthesized by RNA Polymerase III), a type of RNA that accounts for over 50% of the total RNA synthesized in a cell. Pol I consists of 8-14 protein subunits (polypeptides). All 12 subunits have identical or related counterparts in Pol II and Pol III. rDNA transcription is confined to the nucleolus where several hundreds of copies of rRNA genes are present, arranged as tandem head-to-tail repeats. Pol I transcribes one large transcript, encoding an rDNA gene over and over again. This gene encodes the 18S, the 5.8S, and the 28S RNA molecules of the ribosome in eukaryotes. The transcripts are cleaved by snoRNA. The 5S ribosomal RNA is transcribed by Pol III. Because of the simplicity of Pol I transcription, it is the fastest-acting polymerase.

Regulation of rRNA Transcription

The rate of cell growth is directly dependent on the rate of protein synthesis, which, itself, is intricately linked to ribosome synthesis and rRNA transcription. Thus, intracellular signals must coordinate the synthesis of rRNA with that of other components of protein translation. Two specific mechanisms have been identified, ensuring proper control of rRNA synthesis and Pol I-mediated transcription.

Given the large number of rDNA genes (several hundreds) available for transcription, the first mechanism involves adjustments in the number of genes being transcribed at a specific time. In mammalian cells, the number of active rDNA genes varies between cell types and level of differentiation. In general, as a cell become more differentiated, it requires less growth and, therefore, will have a decrease in rRNA synthesis and a decrease in rDNA genes being transcribed. When rRNA synthesis is stimulated, SL1 (selectivity factor 1) will bind to the promoters of rDNA genes that were previously silent, and recruit a pre-initiation complex to which Pol I will bind and start transcription of rRNA.

Changes in rRNA transcription can also occur via changes in the rate of transcription. While the exact mechanism through which Pol I increases its rate of transcription is yet unknown, evidence has shown that rRNA synthesis can increase or decrease without changes in the number of actively transcribed rDNA.

Pol I Transcription Cycle

In the process of transcription (by any polymerase), there are three main stages:

1. Initiation: the construction of the RNA polymerase complex on the gene's promoter with the help of transcription factors
2. Elongation: the actual transcription of the majority of the gene into a corresponding RNA sequence
3. Termination: the cessation of RNA transcription and the disassembly of the RNA polymerase complex.

Initiation

Initiation: the construction of the polymerase complex on the promoter. Pol I requires no TATA box in the promoter, instead relying on a UCS (Upstream Control Sequence).

1. UBF (Upstream Binding Factor) binds the UCS.
2. UCS recruits and binds a protein complex incorporating TBP (TATA Binding Protein) and three TAFs (TBP Associated Factors) called SL1 or TIF-IB. The TBP is forced to bind non-sequence specifically.
3. Rrn3/TIF-IA gets phosphorylated and binds Pol I
4. Pol I binds to the UBF/SL1 complex via Rrn3/TIF-IA, and transcription starts.

Elongation

As Pol I escapes and clears the promoter, UBF and SL1 remain-promoter bound, ready to recruit another Pol I. Indeed, each active rDNA gene can be transcribed multiple times simultaneously, as opposed to Pol II-transcribed genes, which associate with only one complex at a time. While elongation proceeds unimpeded in vitro, it is unclear at this point whether this process happens in a cell, given the presence of nucleosomes. Pol I does seem to transcribe through nucleosomes, either bypassing or disrupting them, perhaps assisted by chromatin-remodelling activities. In addition, UBF might also act as positive feedback, enhancing Pol I elongation through an anti-repressor function. An additional factor, TIF-IC, can also stimulate the overall rate of transcription and suppress pausing of Pol I. As Pol I proceeds along the rDNA, supercoils form both ahead and behind the complex. These are unwound by topoisomerase I or II at regular interval, similar to what is seen in Pol II-mediated transcription.

Elongation is likely to be interrupted at sites of DNA damage. Transcription-coupled repair occurs similarly to Pol II-transcribed genes and require the presence of several DNA repair proteins, such as TFIIH, CSB, and XPG.

Termination

TTF-I binds and bends the termination site at the 3' end of the transcribed region. This will force Pol I to pause. TTF-I, with the help of transcript-release factor PTRF and a T-rich region, will induce Pol I into terminating transcription and dissociating from the DNA and the new transcript. Evidence suggests that termination might be rate-limiting in cases of high rRNA production. TTF-I and PTRF will then indirectly stimulate the reinitiation of transcription by Pol I at the same rDNA gene.

RNA Polymerase II

RNA polymerase II (also called RNAP II and Pol II) is an enzyme found in eukaryotic cells. It catalyzes the transcription of DNA to synthesize precursors of mRNA and most snRNA and microRNA. A 550 kDa complex of 12 subunits, RNAP II is the most studied type of RNA polymerase. A wide range of transcription factors are required for it to bind to its promoters and begin transcription.

Subunits

The eukaryotic core RNA polymerase II was first purified using transcription assays. The purified enzyme has typically 10-12 subunits (12 in humans and yeast) and is incapable of specific promoter recognition. Many subunit-subunit interactions are known.

DNA-directed RNA polymerase II subunit RPB1 is an enzyme that in humans is encoded by the POLR2A gene. RPB1 is the largest subunit of RNA polymerase II. It contains a carboxy terminal domain (CTD) composed of up to 52 heptapeptide repeats (YSPTSPS) that are essential for polymerase activity. In combination with several other polymerase subunits, it forms the DNA binding domain of the polymerase, a groove in which the DNA template is transcribed into RNA. It strongly interacts with RPB8.

RPB2 (POLR2B) is the second largest subunit which in combination with at least two other polymerase subunits forms a structure within the polymerase that maintains contact in the active site of the enzyme between the DNA template and the newly synthesized RNA.

The third largest subunit RPB3 (POLR2C) exists as a heterodimer with POLR2J forming a core subassembly. RPB3 strongly interacts with RPB1-5, 7, 10-12.

RNA polymerase II subunit B4 (RPB4) encoded by the POLR2D gene is the fourth largest subunit and may have a stress protective role.

In humans RPB5 is encoded by the POLR2E gene. Two molecules of this subunit are present in each RNA polymerase II. RPB5 strongly interacts with RPB1, RPB3, and RPB6.

RPB6 (POLR2F) forms a structure with at least two other subunits that stabilizes the transcribing polymerase on the DNA template.

POLR2G encodes RPB7 that may play a role in regulating polymerase function. RPB7 interacts strongly with RPB1 and RPB5.

RPB8 (POLR2H) interacts with subunits RPB1-3, 5, and 7.

The groove in which the DNA template is transcribed into RNA is composed of RPB9 (POLR2I) and RPB1.

RPB10 is the product of gene POLR2L. It interacts with RPB1-3 and 5, and strongly with RPB3.

The RPB11 subunit is itself composed of three subunits in humans: POLR2J (RPB11-a), POLR2J2 (RPB11-b), and POLR2J3 (RPB11-c).

Also interacting with RPB3 is RPB12 (POLR2K).

Assembly

RPB3 is involved in RNA polymerase II assembly. A subcomplex of RPB2 and RPB3 appears soon after subunit synthesis. This complex subsequently interacts with RPB1.

RPB3, RPB5, and RPB7 interact with themselves to form homodimers, and RPB3 and RPB5 together are able to contact all of the other RPB subunits, except RPB9. Only RPB1 strongly binds to RPB5. The RPB1 subunit also contacts RPB7, RPB10, and more weakly but most efficiently with RPB8. Once RPB1 enters the complex, other subunits such as RPB5 and RPB7 can enter, where RPB5 binds to RPB6 and RPB8 and RPB3 brings in RPB10, RPB 11, and RPB12. RPB4 and RPB9 may enter once most of the complex is assembled. RPB4 forms a complex with RPB7.

Kinetics

Enzymes can catalyse up to several million reactions per second. Enzyme rates depend on solution conditions and substrate concentration. Like other enzymes POLR2 has a saturation curve and a maximum velocity (V_{max}). It has a K_m (substrate concentration required for one-half V_{max}) and a k_{cat} (the number of substrate molecules handled by one active site per second. The specificity constant is given by k_{cat}/K_m. The theoretical maximum for the specificity constant is the diffusion limit of about 10^8 to 10^9 (M^{-1} s^{-1}), where every collision of the enzyme with its substrate results in catalysis.

The turnover number for RNA polymerase II is 0.16 s^{-1} subject to concentration. RNA polymerase II switches between inactivated and activated states by translocating back and forth along the DNA. Concentrations of $[NTP]_{eq}$ = 10 μM GTP, 10 μM UTP, 5 μM ATP and 2.5 μM CTP, produce a mean elongation rate, turnover number, of ~1 bp (NTP) s^{-1}.

RNA Polymerase II is inhibited by α-amanitin.

Holoenzyme

RNA polymerase II holoenzyme is a form of eukaryotic RNA polymerase II that is recruited to the promoters of protein-coding genes in living cells. It consists of RNA polymerase II, a subset of general transcription factors, and regulatory proteins known as SRB proteins.

Part of the assembly of the holoenzyme is referred to as the preinitiation complex, because its assembly takes place on the gene promoter before the initiation of transcription. The mediator complex acts as a bridge between RNA polymerase II and the transcription factors.

Control by Chromatin Structure

This is an outline of an example mechanism of yeast cells by which chromatin structure and histone posttranslational modification help regulate and record the transcription of genes by RNA polymerase II.

This pathway gives examples of regulation at these points of transcription:

- Pre-initiation (promotion by Bre1, histone modification)
- Initiation (promotion by TFIIH, Pol II modification AND promotion by COMPASS, histone modification)
- Elongation (promotion by Set2, Histone Modification).

Please note that this refers to various stages of the process as regulatory steps. It has not been proven that they are used for regulation, but is very likely they are. RNA Pol II elongation promoters can be summarised in 3 classes.

1. Drug/sequence-dependent arrest affected factors (Various interfering proteins).
2. Chromatin structure oriented factors (Histone posttranscriptional modifiers, eg HMTs).
3. RNA Pol II catalysis improving factors.

Protein Complexes Involved

Chromatin structure oriented factors:

(HMTs (Histone MethylTransferases)):

COMPASS§† - (COMplex of Proteins ASsociated with Set1) - Methylates lysine 4 of histone H3.

Set2 - Methylates lysine 36 of histone H3.

(interesting irrelevant example: Dot1*‡ - Methylates lysine 79 of histone H3.)

(Other): Bre1 - Ubiquinates (adds ubiquitin to) lysine 123 of histone H2B. Associated with pre-initiation and allowing RNA Pol II binding.

N-terminus

The N-terminus (also known as the amino-terminus, NH_2-terminus, N-terminal end or amine-terminus) refers to the start of a protein or polypeptide terminated by an amino acid with a free amine group ($-NH_2$). The convention for writing peptide sequences is to put the N-terminus on the left and write the sequence from N- to C-terminus. When the protein is translated from messenger RNA, it is created from N-terminus to C-terminus. The N-terminus is the first part of the protein that exits the ribosome during protein biosynthesis. It often contains sequences that act as targeting signals, basically intracellular zip codes, that allow for the protein to be delivered to its designated location within the cell. The targeting signal is usually cleaved off after successful targeting by a processing peptidase. Some proteins are modified posttranslationally.

C-terminus

The C-terminus (also known as the carboxyl-terminus, carboxy-terminus, C-terminal end, or COOH-terminus) of a protein or polypeptide is the end of the amino acid chain terminated by a free carboxyl group (-COOH). The convention for writing peptide sequences is to put the C-terminal end on the right and write the sequence from N- to C-terminus. Each amino acid has a carboxyl group and an amine group, and amino acids link to one another to form a chain by a dehydration reaction by joining the amine group of one amino acid to the carboxyl group of the next. Thus polypeptide chains have an end with an unbound carboxyl group, the C-terminus, and an end with an amine group, the N-terminus. Proteins are naturally synthesized starting from the N-terminus and ending at the C-terminus.

The C-terminus can contain retention signals for protein sorting. The most common ER retention signal is the amino acid sequence -KDEL (or -HDEL) at the C-terminus, which keeps the protein in the endoplasmic reticulum and prevents it from entering the secretory pathway. The C-terminus of proteins can be modified posttranslationally, for example, most commonly by the addition of a lipid anchor to the C-terminus that allows the protein to be inserted into a membrane without having a transmembrane domain. With Pol II, the C-terminus of RPB1 is appended to form the C-terminal domain (CTD).

CTD of RNA Polymerase

The carboxy-terminal domain of RNA polymerase II typically consists of up to 52 repeats of the sequence Tyr-Ser-Pro-Thr-Ser-Pro-Ser. Other proteins often bind the C-terminal domain of RNA polymerase in order to activate polymerase activity. It is the protein domain which is involved in the initiation of DNA transcription, the capping of the RNA transcript, and attachment to the spliceosome for RNA splicing.

RNA Polymerase III

RNA polymerase III (also called Pol III) transcribes DNA to synthesize ribosomal 5S rRNA, tRNA and other small RNAs. The genes transcribed by RNA Pol III fall in the category of "housekeeping" genes whose expression is required in all cell types and most environmental conditions. Therefore the regulation of Pol III transcription is primarily tied to the regulation of cell growth and the cell cycle, thus requiring fewer regulatory proteins than RNA polymerase II.

In the process of transcription (by any polymerase) there are three main stages:

1. Initiation; requiring construction of the RNA polymerase complex on the gene's promoter.
2. Elongation; the writing of the RNA transcript.
3. Termination; the finishing of RNA writing and disassembly of the RNA polymerase complex.

Initiation

Initiation: the construction of the polymerase complex on the promoter. Pol III is unusual (compared to Pol II) requiring no control sequences upstream of the gene, instead normally relying on internal control sequences - sequences within the transcribed section of the gene (although upstream sequences are occasionally seen, eg. U6 snRNA gene has an upstream TATA box as seen in Pol II Promoters).

Class I

Typical stages in 5S rRNA (also termed class I) gene initiation:

1. TFIIIA (Transcription Factor for polymerase III A) binds to the intragenic (lying within the transcribed DNA sequence) 5S rRNA control sequence, the C Block (also termed box C).
2. TFIIIA Serves as a platform that replaces the A and B Blocks for positioning TFIIIC in an orientation with respect to the start site of transcription that is equivalent to what is observed for tRNA genes.
3. Once TFIIIC is bound to the TFIIIA-DNA complex the assembly of TFIIIB proceeds as described for tRNA transcription.

Class II

Typical stages in a tRNA (also termed class II) gene initiation:

1. TFIIIC (Transcription Factor for polymerase III C) binds to two intragenic (lying within the transcribed DNA sequence) control sequences, the A and B Blocks (also termed box A and box B).

2. TFIIIC acts as an assembly factor that positions TFIIIB to bind to DNA at a site centered approximately 26 base pairs upstream of the start site of transcription. TFIIIB (Transcription Factor for polymerase III B), consists of three subunits: TBP (TATA Binding Protein), the Pol II transcription factor TFIIB-related protein, Brf1 (or Brf2 for transcription of a subset of Pol III-transcribed genes in vertebrates) and Bdp1.

3. TFIIIB is the transcription factor that assembles Pol III at the start site of transcription. Once TFIIIB is bound to DNA, TFIIIC is no longer required. TFIIIB also plays an essential role in promoter opening.

TFIIIB remains bound to DNA following initiation of transcription by Pol III (unlike bacterial σ factors and most of the basal transcription factors for Pol II transcription). This leads to a high rate of transcriptional reinitiation of Pol III-transcribed genes.

Class III

Typical stages in a U6 snRNA (also termed class III) gene initiation (documented in vertebrates only):

1. SNAPc (SNRNA Activating Protein complex) (also termed PBP and PTF) binds to the PSE (Proximal Sequence Element) centered approximately 55 base pairs upstream of the start site of transcription. This assembly is greatly stimulated by the Pol II transcription factors Oct1 and STAF that bind to an enhancer-like DSE (Distal Sequence Element) at least 200 base pairs upstream of the start site of transcription. These factors and promoter elements are shared between Pol II and Pol III transcription of snRNA genes.

2. SNAPc acts to assemble TFIIIB at a TATA box centered 26 base pairs upstream of the start site of transcription. It is the presence of a TATA box that specifies that the snRNA gene is transcribed by Pol III rather than Pol II.

3. The TFIIIB for U6 snRNA transcription contains a smaller Brf1 paralogue, Brf2.

4. TFIIIB is the transcription factor that assembles Pol III at the start site of transcription. Sequence conservation predicts that TFIIIB containing Brf2 also plays a role in promoter opening.

Termination

Polymerase III terminates transcription at small polyTs stretch (5-6). In Eukaryotes, a hairpin loop is not required, as it is in prokaryotes.

Transcribed RNAs

The types of RNAs transcribed from RNA polymerase III includes:

- Transfer RNAs
- 5S ribosomal RNA
- U6 spliceosomal RNA
- RNase P and RNase MRP RNA
- 7SL RNA (the RNA component of the signal recognition particle)
- Vault RNAs
- Y RNA
- SINEs (short interspersed repetitive elements)
- 7SK RNA
- Several microRNAs
- Several small nucleolar RNAs
- Several gene regulatory antisense RNAs

6

Analysing Nucleic Acids

In molecular biology, analysis of nucleic acids is commonly performed to determine the average concentrations of DNA or RNA present in a mixture, as well as their purity. Reactions that use nucleic acids often require particular amounts and purity for optimum performance. There are several methods to establish the concentration of a solution of nucleic acids, including spectrophotometric quantification and UV fluorescence in presence of a DNA dye.

Spectrophotometric Analysis

Nucleic acids absorb ultraviolet light in a specific pattern. In a spectrophotometer, a sample is exposed to ultraviolet light at 260 nm, and a photo-detector measures the light that passes through the sample. The more light absorbed by the sample, the higher the nucleic acid concentration in the sample.

Using the Beer Lambert Law it is possible to relate the amount of light absorbed to the concentration of the absorbing molecule. At a wavelength of 260 nm, the average extinction coefficient for double-stranded DNA is 0.020 $(\mu g/ml)^{-1}$ cm^{-1}, for single-stranded DNA and RNA it is 0.027 $(\mu g/ml)^{-1}$ cm^{-1} and for short single-stranded oligonucleotides it is dependent on the length and base composition. Thus, an optical density (or "OD") of 1 corresponds to a concentration of 50 µg/ml for double-stranded DNA. This method of calculation is valid for up to an OD of at least 2. A more accurate extinction coefficient may be needed for oligonucleotides; these can be predicted using the nearest-neighbor model.

Sample Purity

It is common for nucleic acid samples to be contaminated with other molecules (i.e. proteins, organic compounds, other). The ratio of the absorbance at 260 and 280nm ($A_{260/280}$) is used to assess the purity of nucleic acids. For pure DNA, $A_{260/280}$ is ~1.8 and for pure RNA $A_{260/280}$ is ~2.

Protein Contamination and the 260:280 Ratio

The ratio of absorptions at 260nm vs 280nm is commonly used to assess DNA contamination of protein solutions, since proteins (in particular, the aromatic amino acids) absorb light at 280nm.

The reverse, however, is not true — it takes a relatively large amount of protein contamination to significantly affect the 260:280 ratio in a nucleic acid solution.

260:280 ratio has high sensitivity for nucleic acid contamination in protein:

% protein	% nucleic acid	260:280 ratio
100	0	0.57
95	5	1.06
90	10	1.32
70	30	1.73

260:280 ratio lacks sensitivity for protein contamination in nucleic acids:

% nucleic acid	% protein	260:280 ratio
100	0	2.00
95	5	1.99
90	10	1.98
70	30	1.94

This difference is due to the much higher extinction coefficient nucleic acids have at 260nm and 280nm, compared to that of proteins. Because of this, even for relatively high concentrations of protein, the protein contributes relatively little to the 260 and 280 absorbance. While the protein contamination cannot be reliably assessed with a 260:280 ratio, this also means that it contributes little error to DNA quantity estimation.

Other Common Contaminants

- Contamination by phenol, which is commonly used in nucleic acid purification, can significantly throw off quantification estimates. Phenol absorbs with a peak at 270nm and a $A_{260/280}$ of 2. Nucleic acid preparitions uncontaminated by phenol should have a $A_{260/270}$ of around 1.2. Contamination by phenol can significantly contribute to overestimation of DNA concentration.

- Absorption at 230nm can be caused by contamination by phenolate ion, thiocyanates, and other organic compounds. For a pure RNA sample, the $A_{260/230}$ should be around 2, and for a pure DNA sample, the $A_{260/230}$ should be around 1.8.

- Absorption at 330nm and higher indicates particulates contaminating the solution, causing scattering of light in the visible range. The value in a pure nucleic acid sample should be zero.

- Negative values could result if an incorrect solution was used as blank. Alternatively, these values could arise due to fluorescence of a dye in the solution.

Quantification using Fluorescent Dyes

An alternative way to assess DNA concentration is to use measure the fluorescence intensity of dyes that bind to nucleic acids and selectively fluoresce when bound (eg. Ethidium bromide). This method is useful for cases where concentration is too low to accurately assess with spectrophotometry and in cases where contaminants absorbing at 260nm make accurate quantitation by that method impossible.

There are two main ways to approach this. "Spotting" involves placing a sample directly onto an agarose gel or plastic wrap. The fluorescent dye is either present in the agarose gel, or is added in appropriate concentrations to the samples on the plastic film. A set of samples with known concentrations are spotted alongside the sample. The concentration of the unknown sample is then estimated by comparison with the fluorescence of these known concentrations. Alternatively, one may run the sample through an agarose or polyacrylamide gel, alongside some samples of known concentration. As with the spot test, concentration is estimated through comparison of fluorescent intensity with the known samples.

If the sample volumes are large enough to use microplates or cuvettes, the dye-loaded samples can also be quantified with a fluorescence photometer.

Nucleic Acid Methods

Nucleic acid methods are the techniques used to study nucleic acids (DNA and RNA).

Purification

Phenol-chloroform Extraction

Phenol-chloroform extraction is a liquid-liquid extraction technique in biochemistry. It is widely used in molecular biology for isolating DNA, RNA and protein. Equal volumes of a phenol:chloroform mixture and an aqueous sample are mixed, forming a biphasic mixture. This method may take longer than a column-based system such as the silica-based purification, but has higher purity and the advantage of high recovery of RNA: an RNA column is typically unsuitable for purification of short (<200 nucleotides) RNA species, such as siRNA, miRNA and tRNA. Column methods also shear large DNA fragments, which may or may not be a problem depending on downstream applications.

It was originally devised by Piotr Chomczynski and Nicoletta Sacchi and published in 1987 (referred to as *Guanidinium thiocyanate-phenol-chloroform extraction*). It is sold by Sigma-Aldrich by the name TRI Reagent, by Invitrogen under the name TRIzol and by Bioline as Trisure.

How it Works

This method relies on phase separation by centrifugation of a mix of the aqueous sample and a solution containing water-saturated phenol, chloroform and a chaotropic denaturing solution (guanidinium thiocyanate) resulting in an upper aqueous phase and a lower organic phase (mainly chloroform). Nearly all of the RNA is present in the aqueous phase, while DNA and protein partition in the interphase and organic phase, respectively. In a last step, RNA is recovered from the aqueous phase by precipitation with 2-propanol or ethanol. DNA will be located in the aqueous phase in the absence of guanidinium thiocyanate and thus the technique can be used for DNA purification alone.

Guanidinium thiocyanate denatures proteins, including RNases, and separates rRNA from ribosomes, while phenol, isopropanol and water are solvents with poor solubility. In the presence of chloroform or BCP (bromochloropropane), these solvents separate entirely into two phases that are recognized by their colour: a clear, upper aqueous phase (containing the nucleic acids) and a bright pink lower phase (containing the proteins dissolved in phenol and the lipids dissolved in chloroform). Other denaturing chemicals such as 2-mercaptoethanol and sarcosine may also be used. The major downside is that Phenol and chloroform are both hazardous and inconvenient materials, and the extraction is often more laborious, so in recent years many companies now offer alternative ways to isolate DNA.

Reagents

- *Phenol*: The phenol used for biochemistry comes as a water-saturated solution with Tris buffer, as a Tris-buffered 50% phenol, 50% chloroform solution, or as a Tris-buffered 50% phenol, 48% chloroform, 2% isoamyl alcohol solution (sometimes called "25:24:1"). Phenol is naturally somewhat water-soluble, and gives a fuzzy interface, which is sharpened by the presence of chloroform, and the isoamyl alcohol reduces foaming. Most solutions also have an antioxidant, as oxidized phenol damages the nucleic acids. For RNA purification, the pH is kept around pH 4, which retains RNA in the aqueous phase preferentially. For DNA purification, the pH is usually near 7, at which point all nucleic acids are found in the aqueous phase.

- *Chloroform*: Chloroform is stabilized with small quantities of amylene or ethanol, because exposure of pure chloroform to oxygen and ultraviolet light produces phosgene gas. Some chloroform solutions come as pre-made a 96% chloroform, 4% isoamyl alcohol mixtures that can be mixed with an equal volume of phenol to obtain the 25:24:1 solution.

- *Isoamyl alcohol*: Some protocols include isoamyl alcohol as a stabilizing agent, while others do not require it at all.

- Minicolumn purification: Column-based nucleic acid purification is a solid phase extraction method to quickly purify nucleic acids.

This method relies on the fact that the nucleic acid may bind (adsorption) to the solid phase (silica or other) depending on the pH and the salt content of the buffer, which may be a Tris-EDTA (TE) buffer or Phosphate buffer (used in DNA microarray experiments due to the reactive amines).

Therefore, three stages are:

- The sample is added to the column and the nucleic acid binds thanks to the lower pH (relative to the silanol groups on the column) and salt concentration of the binding solution, which may contain buffer, a denaturing agent (such as guanidine hydrochloride), Triton X-100, isopropanol and a pH indicator

- The column is then washed (5 mM KPO4 pH 8.0 or similar, 80% EtOH)
- The column can be eluted with buffer or simply water.

Even prior to the major techniques employed today it was known that in the presence of chaotropic agents, such as sodium iodide or sodium perchlorate, DNA binds to silica, glass particles or to unicellular algae called diatoms which shield their cell walls with silica.

This propriety was used to purify nucleic acid using glass powder or silica beads under alkaline conditions. This was later improved used guanidinium thiocyanate or guanidinium hydrochloride as the chaotropic agent. The use of beads was later changed to minicolumns.

Quantification:

- Abundance in weight: spectroscopic quantification
- Absolute abundance in number: Q-PCR
- high-throughput relative abundance DNA microarray
- high-throughput absolute abundance SAGE
- Size: Gel electrophoresis.

Synthesis:

- De novo: Oligonucleotide synthesis
- Amplification: PCR.

Other:

- Nucleic acid simulations
- DNA sequencing.

The term DNA sequencing refers to sequencing methods for determining the order of the nucleotide bases—adenine, guanine, cytosine, and thymine—in a molecule of DNA.

Knowledge of DNA sequences has become indispensable for basic biological research, other research branches utilizing DNA sequencing, and in numerous applied fields such as diagnostic, biotechnology, forensic biology and biological systematics. The advent of DNA sequencing has significantly accelerated biological research and discovery. The rapid speed of sequencing attained with modern DNA sequencing technology has been instrumental in the sequencing of the human genome, in the Human Genome Project. Related projects, often by scientific collaboration across continents, have generated the complete DNA sequences of many animal, plant, and microbial genomes.

The first DNA sequences were obtained in the early 1970s by academic researchers using laborious methods based on two-dimensional chromatography. Following the development of dye-based sequencing methods with automated analysis, DNA sequencing has become easier and orders of magnitude faster.

History

RNA sequencing was one of the earliest forms of nucleotide sequencing. The major landmark of RNA sequencing is the sequence of the first complete gene and the complete genome of Bacteriophage MS2, identified and published by Walter Fiers and his coworkers at the University of Ghent (Ghent, Belgium), between 1972 and 1976.

Prior to the development of rapid DNA sequencing methods in the early 1970s by Frederick Sanger at the University of Cambridge, in England and Walter Gilbert and Allan Maxam at Harvard, a number of laborious methods were used. For instance, in 1973, Gilbert and Maxam reported the sequence of 24 basepairs using a method known as wandering-spot analysis.

The chain-termination method developed by Sanger and coworkers in 1975 soon became the method of choice, owing to its relative ease and reliability.

Maxam–Gilbert Sequencing

In 1976–1977, Allan Maxam and Walter Gilbert developed a DNA sequencing method based on chemical modification of DNA and subsequent cleavage at specific bases. Although Maxam and Gilbert published their chemical sequencing method two years after the ground-breaking paper of Sanger and Coulson on plus-minus sequencing, Maxam–Gilbert sequencing rapidly became more popular, since purified DNA could be used directly, while the initial Sanger method required that each read start be cloned for production of single-stranded DNA.

However, with the improvement of the chain-termination method, Maxam-Gilbert sequencing has fallen out of favour due to its technical complexity prohibiting its use in standard molecular biology kits, extensive use of hazardous chemicals, and difficulties with scale-up.

The method requires radioactive labelling at one end and purification of the DNA fragment to be sequenced. Chemical treatment generates breaks at a small proportion of one or two of the four nucleotide bases in each of four reactions (G, A+G, C, C+T). Thus a series of labelled fragments is generated, from the radiolabelled end to the first "cut" site in each molecule. The fragments in the four reactions are arranged side by side in gel electrophoresis for size separation. To visualize the fragments, the gel is exposed to X-ray film for autoradiography, yielding a series of dark bands each corresponding to a radiolabelled DNA fragment, from which the sequence may be inferred.

Also sometimes known as "chemical sequencing", this method originated in the study of DNA-protein interactions (footprinting), nucleic acid structure and epigenetic modifications to DNA, and within these it still has important applications.

Chain-termination Methods

Because the chain-terminator method (or Sanger method after its developer Frederick Sanger) is more efficient and uses fewer toxic chemicals and lower amounts of radioactivity than the method of Maxam and Gilbert, it rapidly became the method of choice. The key principle of the Sanger method was the use of dideoxynucleotide triphosphates (ddNTPs) as DNA chain terminators.

The classical chain-termination method requires a single-stranded DNA template, a DNA primer, a DNA polymerase, radioactively or fluorescently labelled nucleotides, and modified nucleotides that terminate DNA strand elongation. The DNA sample is divided into four separate sequencing reactions, containing all four of the standard deoxynucleotides (dATP, dGTP, dCTP and dTTP) and the DNA polymerase. To each reaction is added only one of the four dideoxynucleotides (ddATP, ddGTP, ddCTP, or ddTTP) which are the chain-terminating nucleotides, lacking a 3'-OH group required for the formation of a phosphodiester bond between two nucleotides, thus terminating DNA strand extension and resulting in DNA fragments of varying length.

The newly synthesized and labelled DNA fragments are heat denatured, and separated by size (with a resolution of just one nucleotide) by gel electrophoresis on a denaturing polyacrylamide-urea gel with each of the four reactions run in one of four individual lanes (lanes A, T, G, C); the DNA bands are then visualized by autoradiography or UV light, and the DNA sequence can be directly read off the X-ray film or gel image. In the image on the right, X-ray film was exposed to the gel, and the dark bands correspond to DNA fragments of different lengths. A dark band in a lane indicates a DNA fragment that is the result of chain termination after incorporation of a dideoxynucleotide (ddATP, ddGTP, ddCTP, or ddTTP). The relative positions of the different bands among the four lanes are then used to read (from bottom to top) the DNA sequence.

Technical variations of chain-termination sequencing include tagging with nucleotides containing radioactive phosphorus for radiolabelling, or using a primer labelled at the 5' end with a fluorescent dye. Dye-primer sequencing facilitates reading in an optical system for faster and more economical analysis and automation. The later development by Leroy Hood and coworkers of fluorescently labelled ddNTPs and primers set the stage for automated, high-throughput DNA sequencing. Chain-termination methods have greatly simplified DNA sequencing. For example, chain-termination-based kits are commercially available that contain the reagents needed for sequencing, pre-aliquoted and ready to use. Limitations include non-specific binding of the primer to the DNA, affecting accurate read-out of the DNA sequence, and DNA secondary structures affecting the fidelity of the sequence.

Dye-terminator Sequencing

Dye-terminator sequencing utilizes labelling of the chain terminator ddNTPs, which permits sequencing in a single reaction, rather than four reactions as in the labelled-primer method. In dye-terminator sequencing, each of the four dideoxynucleotide chain terminators is labelled with fluorescent dyes, each of which emit light at different wavelengths. Owing to its greater expediency and speed, dye-terminator sequencing is now the mainstay in automated sequencing. Its limitations include dye effects due to differences in the incorporation of the dye-labelled chain terminators into the DNA fragment, resulting in unequal peak heights and shapes in the electronic DNA sequence trace chromatogram after capillary electrophoresis.

This problem has been addressed with the use of modified DNA polymerase enzyme systems and dyes that minimize incorporation variability, as well as methods for eliminating "dye blobs". The dye-terminator sequencing method, along with automated high-throughput DNA sequence analyzers, is now being used for the vast majority of sequencing projects.

Challenges

Common challenges of DNA sequencing include poor quality in the first 15–40 bases of the sequence and deteriorating quality of sequencing traces after 700–900 bases. Base calling software typically gives an estimate of quality to aid in quality trimming. In cases where DNA fragments are cloned before sequencing, the resulting sequence may contain parts of the cloning vector. In contrast, PCR-based cloning and emerging sequencing technologies based on pyrosequencing often avoid using cloning vectors. Recently, one-step Sanger sequencing (combined amplification and sequencing) methods such as Ampliseq and SeqSharp have been developed that allow rapid sequencing of target genes without cloning or prior amplification.

Current methods can directly sequence only relatively short (300–1000 nucleotides long) DNA fragments in a single reaction. The main obstacle to sequencing DNA fragments above this size limit is insufficient power of separation for resolving large DNA fragments that differ in length by only one nucleotide.

Automation and Sample Preparation

Automated DNA-sequencing instruments (DNA sequencers) can sequence up to 384 DNA samples in a single batch (run) in up to 24 runs a day. DNA sequencers carry out capillary electrophoresis for size separation, detection and recording of dye fluorescence, and data output as fluorescent peak trace chromatograms. Sequencing reactions by thermocycling, cleanup and re-suspension in a buffer solution before loading onto the sequencer are performed separately. A number of commercial and non-commercial software packages can trim low-quality DNA traces automatically. These programs score the quality of each peak and remove low-quality base peaks (generally located at the ends of the sequence). The accuracy of such algorithms is below visual examination by a human operator, but sufficient for automated processing of large sequence data sets.

Amplification and Clonal Selection

Large-scale sequencing aims at sequencing very long DNA pieces, such as whole chromosomes. Common approaches consist of cutting (with restriction enzymes) or shearing (with mechanical forces) large DNA fragments into shorter DNA fragments. The fragmented DNA is cloned into a DNA vector, and amplified in *Escherichia coli*. Short DNA fragments purified from individual bacterial colonies are individually sequenced and assembled electronically into one long, contiguous sequence.

This method does not require any pre-existing information about the sequence of the DNA and is referred to as *de novo* sequencing.

Gaps in the assembled sequence may be filled by primer walking. The different strategies have different tradeoffs in speed and accuracy; *shotgun methods* are often used for sequencing large genomes, but its assembly is complex and difficult, particularly with sequence repeats often causing gaps in genome assembly.

Most sequencing approaches use an *in vitro* cloning step to amplify individual DNA molecules, because their molecular detection methods are not sensitive enough for single molecule sequencing. Emulsion PCR isolates individual DNA molecules along with primer-coated beads in aqueous droplets within an oil phase. Polymerase chain reaction (PCR) then coats each bead with clonal copies of the DNA molecule followed by immobilization for later sequencing. Emulsion PCR is used in the methods by Marguilis et al. (commercialized by 454 Life Sciences), Shendure and Porreca et al. (also known as "Polony sequencing") and SOLiD sequencing, (developed by Agencourt, now Applied Biosystems).

Another method for *in vitro* clonal amplification is *bridge PCR*, where fragments are amplified upon primers attached to a solid surface, used in the Illumina Genome Analyzer. The single-molecule method developed by Stephen Quake's laboratory (later commercialized by Helicos) is an exception: it uses bright fluorophores and laser excitation to detect pyrosequencing events from individual DNA molecules fixed to a surface, eliminating the need for molecular amplification.

High-throughput Sequencing

The high demand for low-cost sequencing has driven the development of high-throughput sequencing technologies that parallelize the sequencing process, producing thousands or millions of sequences at once. High-throughput sequencing technologies are intended to lower the cost of DNA sequencing beyond what is possible with standard dye-terminator methods.

Lynx Therapeutics' Massively Parallel Signature Sequencing (MPSS)

The first of the "next-generation" sequencing technologies, MPSS was developed in 1990s at Lynx Therapeutics, a company founded in 1992 by Sidney Brenner and Sam Eletr. MPSS was a bead-based method that used a complex approach of adapter ligation followed by adapter decoding, reading the sequence in increments of four nucleotides; this method made it susceptible to sequence-specific bias or loss of specific sequences. Because the technology was so complex,

MPSS was only performed 'in-house' by Lynx Therapeutics and no machines were sold; when the merger with Solexa later lead to the development of sequencing-by-synthesis, a more simple approach with numerous advantages, MPSS became obsolete. However, the essential properties of the MPSS output were typical of later "next-gen" data types, including hundreds of thousands of short DNA sequences. In the case of MPSS, these were typically used for sequencing cDNA for measurements of gene expression levels. Lynx Therapeutics merged with Solexa in 2004, and this company was later purchased by Illumina.

454 Pyrosequencing

A parallelized version of pyrosequencing was developed by 454 Life Sciences. The method amplifies DNA inside water droplets in an oil solution (emulsion PCR), with each droplet containing a single DNA template attached to a single primer-coated bead that then forms a clonal colony. The sequencing machine contains many picolitre-volume wells each containing a single bead and sequencing enzymes. Pyrosequencing uses luciferase to generate light for detection of the individual nucleotides added to the nascent DNA, and the combined data are used to generate sequence read-outs. This technology provides intermediate read length and price per base compared to Sanger sequencing on one end and Solexa and SOLiD on the other. 454 Life Sciences has since been acquired by Roche Diagnostics.

Illumina (Solexa) Sequencing

Solexa, now part of Illumina developed a sequencing technology based on reversible dye-terminators. DNA molecules are first attached to primers on a slide and amplified so that local clonal colonies are formed (bridge amplification). One type of nucleotide at a time is then added, and non-incorporated nucleotides are washed away. Unlike pyrosequencing, the DNA can only be extended one nucleotide at a time. A camera takes images of the fluorescently labelled nucleotides and the dye is chemically removed from the DNA, allowing a next cycle.

SOLiD Sequencing

Applied Biosystems' SOLiD technology employs sequencing by ligation. Here, a pool of all possible oligonucleotides of a fixed length are labelled according to the sequenced position. Oligonucleotides are annealed and ligated; the preferential ligation by DNA ligase for matching sequences results in a signal informative of the nucleotide at that position. Before sequencing, the DNA is amplified by emulsion PCR. The resulting bead, each containing only copies of the same DNA molecule, are deposited on a glass slide. The result is sequences of quantities and lengths comparable to Illumina sequencing.

Future Methods

Sequencing by hybridization is a non-enzymatic method that uses a DNA microarray. A single pool of DNA whose sequence is to be determined is fluorescently labelled and hybridized to an array containing known sequences. Strong hybridization signals from a given spot on the array identifies its sequence in the DNA being sequenced. Mass spectrometry may be used to determine mass differences between DNA fragments produced in chain-termination reactions.

DNA sequencing methods currently under development include labelling the DNA polymerase, reading the sequence as a DNA strand transits through nanopores, and microscopy-based techniques, such as AFM or electron microscopy that are used to identify the positions of individual nucleotides within long DNA fragments (>5,000 bp) by nucleotide labelling with heavier elements (e.g., halogens) for visual detection and recording.

In microfluidic Sanger sequencing the entire thermocycling amplification of DNA fragments as well as their separation by electrophoresis is done on a single glass wafer (approximately 10 cm in diameter) thus reducing the reagent usage as well as cost. In some instances researchers have shown that they can increase the throughput of conventional sequencing through the use of microchips. Research will still need to be done in order to make this use of technology effective. In October 2006, the X Prize Foundation established an initiative to promote the development of full genome sequencing technologies, called the Archon X Prize, intending to award $10 million to "the first Team that can build a device and use it to sequence 100 human genomes within 10 days or less, with an accuracy of no more than one error in every 100,000 bases sequenced, with sequences accurately covering at least 98% of the genome, and at a recurring cost of no more than $10,000 (US) per genome."

Bisulfite Sequencing

Bisulfite sequencing is the use of bisulfite treatment of DNA to determine its pattern of methylation. DNA methylation was the first discovered epigenetic mark, and remains the most studied. In animals it predominantly involves the addition of a methyl group to the carbon-5 position of cytosine residues of the dinucleotide CpG, and is implicated in repression of transcriptional activity.

Treatment of DNA with bisulfite converts cytosine residues to uracil, but leaves 5-methylcytosine residues unaffected. Thus, bisulfite treatment introduces specific changes in the DNA sequence that depend on the methylation status of individual cytosine residues, yielding single- nucleotide resolution information about the methylation status of a segment of DNA. Various analyses can be performed on the altered sequence to retrieve this information. The objective of this analysis is therefore reduced to differentiating between single nucleotide polymorphisms (cytosines and thymidine) resulting from bisulfite conversion.

Methods

Bisulfite sequencing applies routine sequencing methods on bisulfite-treated genomic DNA to determine methylation status at CpG dinucleotides. Other non-sequencing strategies are also employed to interrogate the methylation at specific loci or at a genome-wide level. All strategies assume that bisulfite-induced conversion of unmethylated cytosines to uracil is complete, and this serves as the basis of all subsequent techniques. Ideally, the method used would determine the methylation status separately for each allele. Alternative methods to bisulfite sequencing include methylation-specific restriction analysis and methylated DNA immunoprecipitation (MeDIP).

Methodologies to analyze bisulfite-treated DNA are continuously being developed. To summarize these rapidly evolving methologies, numerous review articles have been written.

The methodologies can be generally divided into strategies based on methylation-specific PCR (MSP), and strategies employing polymerase chain reaction (PCR) performed under non-methylation-specific conditions. Microarray-based methods use PCR based on non-methylation-specific conditions also.

Non-methylation-specific PCR based Methods

Direct Sequencing

The first reported method of methylation analysis using bisulfite-treated DNA utilized PCR and standard dideoxynucleotide DNA sequencing to directly determine the nucleotides resistant to bisulfite conversion. Primers are designed to be strand-specific as well as bisulfite-specific (i.e., primers containing non-CpG cytosines such that they are not complementary to non-bisulfite-treated DNA), flanking (but not involving) the methylation site of interest.

Therefore, it will amplify both methylated and unmethylated sequences, in contrast to methylation-specific PCR. All sites of unmethylated cytosines are displayed as thymines in the resulting amplified sequence of the sense strand, and as adenines in the amplified antisense strand. This technique required cloning of the PCR product prior to sequencing for adequate sensitivity, and therefore was a very labour-intensive method unsuitable for higher throughput. Alternatively, nested PCR methods can be used to enhance the product for sequencing.

All subsequent DNA methylation analysis techniques using bisulfite-treated DNA is based on this report by Frommer et al.. Although most other modalities are not true sequencing-based techniques, the term "bisulfite sequencing" is often used to describe bisulfite-conversion DNA methylation analysis techniques in general.

Pyrosequencing

Pyrosequencing has also been used to analyze bisulfite-treated DNA without using methylation-specific PCR. Following PCR amplification of the region of interest, Pyrosequencing is used to determine the bisulfite-converted sequence of specific CpG sites in the region. The ratio of C-to-T at individual sites can be determined quantitatively based on the amount of C and T incorporation during the sequence extension. The main limitation of this method is the cost of the technology. However, Pyrosequencing does well allow for extension to high-throughput screening methods.

A further improvement to this technique was recently described by Wong et al., which uses allele-specific primers that incorporate single-nucleotide polymorphisms into the sequence of the sequencing primer, thus allowing for separate analysis of maternal and paternal alleles. This technique is of particular usefulness for genomic imprinting analysis.

Methylation-sensitive Single-strand Conformation analysis (MS-SSCA)

This method is based on the single-strand conformation polymorphism analysis (SSCA) method developed for single-nucleotide polymorphism (SNP) analysis. SSCA differentiates between single-stranded DNA fragments of identical size but distinct sequence based on differential migration in non-denaturing electrophoresis. In MS-SSCA, this is used to distinguish between bisulfite-treated, PCR-amplified regions containing the CpG sites of interest. Although SSCA lacks sensitivity when only a single nucleotide difference is present, bisulfite treatment frequently makes a number of C-to-T conversions in most regions of interest, and the resulting sensitivity approaches 100%. MS-SSCA also provides semi-quantitative analysis of the degree of DNA methylation based on the ratio of band intensities. However, this method is designed to assess all CpG sites as a whole in the region of interest rather than individual methylation sites.

High Resolution Melting Analysis (HRM)

A further method to differentiate converted from unconverted bisulfite-treated DNA is using high-resolution melting analysis (HRM), a real-time PCR-based technique initially designed to distinguish SNPs. The PCR amplicons are analysed directly by temperature ramping and resulting liberation of an intercalating fluorescent dye during melting. The degree of methylation, as represented by the C-to-T content in the amplicon, determines the rapidity of melting and consequent release of the dye. This method allows direct quantitation in a single-tube assay, but assesses methylation in the amplified region as a whole rather than at specific CpG sites.

Methylation-sensitive Single-nucleotide Primer Extension (MS-SnuPE)

MS-SnuPE employs the primer extension method initially designed for analysing single-nucleotide polymorphisms. DNA is bisulfite-converted, and bisulfite-specific primers are annealed to the sequence up to the base pair immediately before the CpG of interest. The primer is allowed to extend one base pair into the C (or T) using DNA polymerase terminating dideoxynucleotides, and the ratio of C to T is determined quantitatively.

A number of methods can be used to determine this C:T ratio. At the beginning, MS-SnuPE relied on radioactive ddNTPs as the reporter of the primer extension. Fluorescence-based methods or Pyrosequencing can also be used. However, matrix-assisted laser desorption ionization/time-of-flight (MALDI-TOF) mass spectrometry analysis to differentiate between the two polymorphic primer extension products can be used, in essence, based on the GOOD assay designed for SNP genotyping. Ion pair reverse-phase high-performance liquid chromatography (IP-RP-HPLC) has also been used to distinguish primer extension products.

Base-specific Cleavage/MALDI-TOF

A recently described method by Ehrich et al. further takes advantage of bisulfite-conversions by adding a base-specific cleavage step to enhance the information gained from the nucleotide changes. By first using in vitro transcription of the region of interest into RNA (by adding an RNA polymerase promoter site to the PCR primer in the initial amplification), RNase A can be used to cleave the RNA transcript at base-specific sites. As RNase A cleaves RNA specifically at cytosine and uracil ribonucleotides, base-specificity is achieved by adding incorporating cleavage-resistant dTTP when cytosine-specific (C-specific) cleavage is desired, and incorporating dCTP when uracil-specific (U-specific) cleavage is desired. The cleaved fragments can then be analysed by MALDI-TOF. Bisulfite treatment results in either introduction/removal of cleavage sites by C-to-U conversions or shift in fragment mass by G-to-A conversions in the amplified reverse strand. C-specific cleavage will cut specifically at all methylated CpG sites. By analysing the sizes of the resulting fragments, it is possible to determine the specific pattern of DNA methylation of CpG sites within the region, rather than determining the extent of methylation of the region as a whole. This method demonstrated efficacy for high-throughput screening, allowing for interrogation of numerous CpG sites in multiple tissues in a cost-efficient manner.

Methylation-specific PCR (MSP)

This alternative method of methylation analysis also uses bisulfite-treated DNA but avoids the need to sequence the area of interest. Instead, primer pairs are designed themselves to be "methylated-specific" by including sequences complementing only unconverted 5-methylcytosines, or, on the converse, "unmethylated-specific", complementing thymines converted from unmethylated cytosines. Methylation is determined by the ability of the specific primer to achieve amplification. This method is particularly useful to interrogate CpG islands with possibly high methylation density, as increased numbers of CpG pairs in the primer increase the specificity of the assay. Placing the CpG pair at the 3'-end of the primer also improves the sensitivity. The initial report using MSP described sufficient sensitivity to detect methylation of 0.1% of alleles. In general, MSP and its related protocols are considered to be the most sensitive when interrogating the methylation status at a specific locus.

The MethyLight method is based on MSP, but provides a quantitative analysis using real-time PCR. Methylated-specific primers are used, and a methylated-specific fluorescence reporter probe is also used that anneals to the amplified region. In alternative fashion, the primers or probe can be designed without methylation specificity if discrimination is needed between the CpG pairs within the involved sequences. Quantitation is made in reference to a methylated reference DNA. A modification to this protocol to increase the specificity of the PCR for successfully bisulfite-converted DNA (ConLight-MSP) uses an additional probe to bisulfite-unconverted DNA to quantify this non-specific amplification.

Further methodology using MSP-amplified DNA analyzes the products using melting-curve analysis (Mc-MSP). This method amplifies bisulfite-converted DNA with both methylated-specific and unmethylated-specific primers, and determines the quantitative ratio of the two products by comparing the differential peaks generated in a melting-curve analysis. A high-resolution melting analysis method that uses both real-time quantification and melting analysis has been introduced, in particular, for sensitive detection of low-level methylation

Microarray-based Methods

Microarray-based methods are a logical extension of the technologies available to analyze bisulfite-treated DNA to allow for genome-wide analysis of methylation. Oligonucleotide microarrays are designed using oligonucleotide pairs targeting CpG sites of interest, with one complementary to the unaltered methylated sequence, and the other to the C-to-U-converted unmethylated sequence. The oligonucleotides are also bisulfite-specific to prevent binding to DNA incompletely converted by bisulfite. The Illumina Methylation Assay is one such assay that applies the bisulfite sequencing technology on a microarray level to generate genome-wide methylation data.

Limitations of Incomplete Conversion

Bisulphite sequencing relies on the conversion of every single unmethylated cytosine residue to uracil. If conversion is incomplete, the subsequent analysis will incorrectly interpret the unconverted unmethylated cytosines as methylated cytosines, resulting in false positive results for methylation. Only cytosines in single-stranded DNA are susceptible to attack by bisulphite, therefore denaturation of the DNA undergoing analysis is critical. It is important to ensure that reaction parameters such as temperature and salt concentration are suitable to maintain the DNA in a single-stranded conformation and allow for complete conversion. Embedding the DNA in agarose gel has been reported to improve the rate of conversion by keeping strands of DNA physically separate.

Degradation of DNA during Bisulphite Treatment

A major challenge in bisulphite sequencing is the degradation of DNA that takes place concurrently with the conversion. The conditions necessary for complete conversion, such as long incubation times, elevated temperature, and high bisulphite concentration, can lead to the degradation of about 90% of the incubated DNA. Given that the starting amount of DNA is often limited, such extensive degradation can be problematic. The degradation occurs as depurinations resulting in random strand breaks. Therefore the longer the desired PCR amplicon, the more limited the number of intact template molecules will likely be. This could lead to the failure of the PCR amplification, or the loss of quantitatively accurate information on methylation levels resulting from the limited sampling of template molecules. Thus, it is important to assess the amount of DNA degradation resulting from the reaction conditions employed, and consider how this will affect the desired amplicon. Techniques can also be used to minimize DNA degradation, such as cycling the incubation temperature.

Other Concerns

A potentially significant problem following bisulphite treatment is incomplete desulfonation of pyrimidine residues due to inadequate alkalization of the solution. This may inhibit some DNA polymerases, rendering subsequent PCR difficult. However, this situation can be avoided by monitoring the pH of the solution to ensure that desulphonation will be complete. A final concern is that bisulphite treatment greatly reduces the level of complexity in the sample, which can be problematic if multiple PCR reactions are to be performed (2006). Primer design is more difficult, and inappropriate cross-hybridization is more frequent.

Applications: Genome-wide Methylation Analysis

The advances in bisulfite sequencing have led to the possibility of applying them at a genome-wide scale, where, previously, global measure of DNA methylation was feasible only using other techniques, such as Restriction landmark genomic scanning. The mapping of the human epigenome is seen by many scientists as the logical follow-up to the completion of the Human Genome Project. This epigenomic information will be important in understanding how the function of the genetic sequence is implemented and regulated. Since the epigenome is less stable than the genome, it is thought to be important in gene-environment interactions.

Epigenomic mapping is inherently more complex than genome sequencing, however, since the epigenome is much more variable than the genome. While an individual has only one genome, one's epigenome varies with age, differs between tissues, is altered by environmental factors, and shows aberrations in diseases. Such rich epigenomic mapping, however, representing different ages, tissue types, and disease states, would yield valuable information on the normal function of epigenetic marks as well as the mechanisms leading to aging and disease.

Direct benefits of epigenomic mapping include probable advances in cloning technology. It is believed that failures to produce cloned animals with normal viability and lifespan result from inappropriate patterns of epigenetic marks. Also, aberrant methylation patterns are well characterized in many cancers. Global hypomethylation results in decreased genomic stability, while local hypermethylation of tumour suppressor gene promoters often accounts for their loss of function. Specific patterns of methylation are indicative of specific cancer types, have prognostic value, and can help to guide the best course of treatment.

Large-scale epigenome mapping efforts are under way around the world and have been organized under the Human Epigenome Project. This is based on a multi-tiered strategy, whereby bisulfite sequencing is used to obtain high-resolution methylation profiles for a limited number of reference epigenomes, while less thorough analysis is performed on a wider spectrum of samples. This approach is intended to maximize the insight gained from a given amount of resources, as high-resolution genome-wide mapping remains a costly undertaking.

Expression Cloning

Expression cloning is a technique in DNA cloning that uses expression vectors to generate a library of clones, with each clone expressing one protein. This *expression library* is then screened for the property of interest and clones of interest recovered for further analysis. An example would be using an expression library to isolate genes that could confer antibiotic resistance.

Expression Vectors

Expression vectors are a specialized type of cloning vector in which the transcriptional and translational signals needed for the regulation of the gene of interest are included in the cloning vector. The transcriptional and translational signals may be synthetically created to make the expression of the gene of interest easier to regulate.

Purpose

Usually the ultimate aim of expression cloning is to produce large quantities of specific proteins. To this end, a bacterial expression clone may include a ribosome binding site (Shine-Dalgarno sequence) to enhance translation of the gene of interest's mRNA, a transcription termination sequence, or, in eukaryotes, specific sequences to promote the post-translational modification of the protein product.

Southern Blot

A Southern blot is a method routinely used in molecular biology for detection of a specific DNA sequence in DNA samples. Southern blotting combines transfer of electrophoresis-separated DNA fragments to a filter membrane and subsequent fragment detection by probe hybridization. The method is named after its inventor, the British biologist Edwin Southern. Other blotting methods (i.e., Western blot, Northern blot, Eastern blot, Southwestern blot) that employ similar principles, but using RNA or protein, have later been named in reference to Edwin Southern's name. As the technique was eponymously named, Southern blot is capitalized as is conventional for proper nouns. The names for other blotting methods follow this convention by analogy.

Northern Blot

The northern blot is a technique used in molecular biology research to study gene expression by detection of RNA (or isolated mRNA) in a sample.

With northern blotting it is possible to observe cellular control over structure and function by determining the particular gene expression levels during differentiation, morphogenesis, as well as abnormal or diseased conditions. Northern blotting involves the use of electrophoresis to separate RNA samples by size, and detection with a hybridization probe complementary to part of or the entire target sequence. The term 'northern blot' actually refers specifically to the capillary transfer of RNA from the electrophoresis gel to the blotting membrane, however the entire process is commonly referred to as northern blotting. The northern blot technique was developed in 1977 by James Alwine, David Kemp, and George Stark at Stanford University. Northern blotting takes its name from its similarity to the first blotting technique, the Southern blot, named for biologist Edwin Southern. The major difference is that RNA, rather than DNA, is analysed in the Northern blot.

Procedure

A general blotting procedure starts with extraction of total RNA from a homogenized tissue sample. The mRNA can then be isolated through the use of oligo (dT) cellulose chromatography to maintain only those RNAs with a poly(A) tail. RNA samples are then separated by gel electrophoresis. Since the gels are fragile and the probes are unable to enter the matrix, the RNA samples, now separated by size, are transferred to a nylon membrane through a capillary or vacuum blotting system.

A nylon membrane with a positive charge is the most effective for use in northern blotting since the negatively charged nucleic acids have a high affinity for them. The transfer buffer used for the blotting usually contains formamide because it lowers the annealing temperature of the probe-RNA interaction preventing RNA degradation by high temperatures. Once the RNA has been transferred to the membrane it is immobilized through covalent linkage to the membrane by UV light or heat. After a probe has been labelled, it is hybridized to the RNA on the membrane. Experimental conditions that can affect the efficiency and specificity of hybridization include ionic strength, viscosity, duplex length, mismatched base pairs, and base composition. The membrane is washed to ensure that the probe has bound specifically and to avoid background signals from arising. The hybrid signals are then detected by X-ray film and can be quantified by densitometry. To create controls for comparison in a northern blot, samples not displaying the gene product of interest can be used after determination by microarrays or RT-PCR.

Gels

The RNA samples are most commonly separated on agarose gels containing formaldehyde as a denaturing agent for the RNA to limit secondary structure. The gels can be stained with ethidium bromide

(EtBr) and viewed under UV light to observe the quality and quantity of RNA before blotting. Polyacrylamide gel electrophoeresis with urea can also be used in RNA separation but it is most commonly used for fragmented RNA or microRNAs. An RNA ladder is often run alongside the samples on an electrophoresis gel to observe the size of fragments obtained but in total RNA samples the ribosomal subunits can act as size markers. Since the large ribosomal subunit is 28S (approximately 5kb) and the small ribosomal subunit is 18S (approximately 2kb) two prominent bands will appear on the gel, the larger at close to twice the intensity of the smaller.

Probes

Probes for northern blotting are composed of nucleic acids with a complementary sequence to all or part of the RNA of interest, they can be DNA, RNA, or oligonucleotides with a minimum of 25 complementary bases to the target sequence. RNA probes (riboprobes) that are transcribed in vitro are able to withstand more rigorous washing steps preventing some of the background noise. Commonly cDNA is created with labelled primers for the RNA sequence of interest to act as the probe in the northern blot. The probes need to be labelled either with radioactive isotopes or with chemiluminescence in which alkaline phosphatase or horseradish peroxidase breakdown chemiluminescent substrates producing a detectable emission of light. The chemiluminescent labelling can occur in two ways: either the probe is attached to the enzyme, or the probe is labelled with a ligand (e.g. biotin) for which the antibody (e.g. avidin or streptavidin) is attached to the enzyme. X-ray film can detect both the radioactive and chemiluminescent signals and many researchers prefer the chemiluminescent signals because they are faster, more sensitive, and reduce the health hazards that go along with radioactive labels. The same membrane can be probed up to five times without a significant loss of the target RNA.

Applications

Northern blotting allows one to observe a particular gene's expression pattern between tissues, organs, developmental stages, environmental stress levels, pathogen infection, and over the course of treatment. The technique has been used to show overexpression of oncogenes and downregulation of tumour-suppressor genes in cancerous cells when compared to 'normal' tissue, as well as the gene expression in the rejection of transplanted organs. If an upregulated gene is observed by an abundance of mRNA on the northern blot the sample can then be sequenced to determine if the gene is known to researchers or if it is a novel finding. The expression patterns obtained under given conditions can provide insight into the function of that gene. Since the RNA is first separated by size, if only one probe type is used variance in the level of each band on the membrane can provide insight into the size of the product, suggesting alternative splice products of the same gene or repetitive sequence motifs. The variance in size of a gene product can also indicate deletions or errors in transcript processing, by altering the probe target used along the known sequence it is possible to determine which region of the RNA is missing.

BlotBase is an online database publishing northern blots. BlotBase has over 700 published northern blots of human and mouse samples, in over 650 genes across more than 25 different tissue types. Northern blots can be searched by a blot ID, paper reference, gene identifier, or by tissue. The results of a search provide the blot ID, species, tissue, gene, expression level, blot image (if available), and links to the publication that the work originated from. This new database provides sharing of information between members of the science community that was not previously seen in northern blotting as it was in sequence analysis, genome determination, protein structure, etc.

Disadvantages and Advantages

Analysis of gene expression can be done by several different methods including RT-PCR, RNase protection assays, microarrays, serial analysis of gene expression (SAGE), as well as northern blotting. Microarrays are quite commonly used and are usually consistent with data obtained from northern blots; however, at times northern blotting is able to detect small changes in gene expression that microarrays cannot. The advantage that microarrays have over northern blots is that thousands of genes can be visualized at a time, while northern blotting is usually looking at one or a small number of genes.

A problem in northern blotting is often sample degradation by RNases (both endogenous to the sample and through environmental contamination), which can be avoided by proper sterilization of glassware and the use of RNase inhibitors such as DEPC (diethylpyrocarbonate). The chemicals used in most northern blots can be a risk to the researcher, since formaldehyde, radioactive material, ethidium bromide, DEPC, and UV light are all harmful under certain exposures. Compared to RT-PCR, northern blotting has a low sensitivity, but it also has a high specificity which is important to reduce false positive results.

The advantages of using northern blotting include the detection of RNA size, the observation of alternate splice products, the use of probes with partial homology, the quality and quantity of RNA can be measured on the gel prior to blotting, and the membranes can be stored and reprobed for years after blotting.

Reverse Northern Blot

A variant of the procedure known as the reverse northern blot is occasionally used. In this procedure, the substrate nucleic acid (that is affixed to the membrane) is a collection of isolated DNA fragments, and the probe is RNA extracted from a tissue and radioactively labelled.

The use of DNA microarrays that have come into widespread use in the late 1990s and early 2000s is more akin to the reverse procedure, in that they involve the use of isolated DNA fragments affixed to a substrate, and hybridization with a probe made from cellular RNA. Thus the reverse procedure, though originally uncommon, enabled northern analysis to evolve into gene expression profiling, in which many (possibly all) of the genes in an organism may have their expression monitored.

Sucrose Gradient Centrifugation

Sucrose gradient centrifugation is a type of centrifugation often used to purify enveloped viruses (with densities 1.1-1.2 g/cm^3) and ribosomes, and also to separate cell organelles from crude cellular extracts. This method is also used to purify exosomes.

Equilibrium Centrifugation

Typically, a sucrose density gradient is created by gently overlaying lower concentrations of sucrose on higher concentrations in a centrifuge tube. For example, a sucrose gradient may consist of layers extending from 70% sucrose to 20% sucrose in 10% increments (though this is highly variable depending on sample to be purified). The sample containing the particles of interest is placed on top of the gradient and centrifuged at forces in excess of 150,000 x g. The particles travel through the gradient until they reach the point in the gradient at which their density matches that of the surrounding sucrose. This fraction can then be removed and subjected to further analysis. A similar technique is sucrose cushion centrifugation, in which a particle mixture is pelleted through a 20% sucrose layer, coming to rest at the interface with a 70% solution.

This allows concentration of particles from a sample. Unlike standard centrifugation, which in effect crushes the particles against the bottom of the centrifuge tube, the sucrose cushion method causes no mechanical stress and allows the collection of morphologically intact particles.

Non-equilibrium Centrifugation

This is very similar to the equilibrium form, but the experiment is only run until a particular point. Then the sucrose is eluted from the bottom of the tube.

Radioactivity in Biological Research

Radioactivity can be used in life sciences as a radiolabel to visualise components or target molecules in a biological system. Some radionuclei are synthesised in particle accelerators and have short half-lives, giving them high maximum theoretical specific activities. This lowers the detection time compared to radionuclei with longer half-lives, such as carbon-14. In some applications they have been substituted by fluorescent dyes.

Examples of Radionuclei

- Tritium (hydrogen-3) is a very low energy emitter that can be used to label proteins, nucleic acids, drugs and toxins, but requires a tritium-specific film or a tritium-specific phosphor screen. In a liquid scintillation assay (LSA), the efficiency is 20–50%, depending on the scintillation cocktail used. The maximum theoretical specific activity of tritium is 28.8 Ci/mmol (1.066 PBq/mol). However, there is often more than one tritium atom per molecule: for example, tritiated UTP is sold by most suppliers with carbons 5 and 6 each bonded to a tritium atom. C-14, S-35 and P-33 have similar emission energies. P-32 and I-125 are higher energy emitters -> inaccurate, see beta vs gamma radiation.
- Carbon-14 has a long half-life of 5,730±40 years. Its maximum specific activity is 0.0624 Ci/mmol (2.31 TBq/mol). It is used in applications such as radiometric dating or drug tests.

- Sodium-22 and chlorine-36 are commonly used to study ion transporters. However, sodium-22 is hard to screen off and chlorine-36, with a half-life of 300,000 years, has low activity.

- Sulfur-35 is used to label proteins and nucleic acids. Cysteine is an amino acid containing a thiol group which can be labelled by S-35. For nucleotides that do not contain a sulfur group, the oxygen on one of the phosphate groups can be substituted with a sulfur. This thiophosphate acts the same as a normal phosphate group, although there is a slight bias against it by most polymerases. The maximum theoretical specific activity is 1,494 Ci/mmol (55.28 PBq/mol).

- Phosphorus-33 is used to label nucleotides. It is less energetic than P-32 and does not require protection with plexi glass. A disadvantage is its higher cost compared to P-32, as most of the bombarded P-31 will have acquired only one neutron, while only some will have acquired two or more. Its maximum specific activity is 5,118 Ci/mmol (189.4 PBq/mol).

- Phosphorus-32 is widely used for labelling nucleic acids and phosphoproteins. It has the highest emission energy (1.7 MeV) of all common research radioisotopes. This is a major advantage in experiments for which sensitivity is a primary consideration, such as titrations of very strong interactions (i.e., very low dissociation constant), footprinting experiments, and detection of low-abundance phosphorylated species. 32P is also relatively inexpensive. Because of its high energy, however, a number of safety and administrative controls are required (e.g., acrylic glass). The half-life of 32P is 14.2 days, and its maximum specific activity is 9131 Ci/mmol.

- Iodine-125 is commonly used for labelling proteins, usually at tyrosine residues. Unbound iodine is volatile and must be handled in a fume hood. Its maximum specific activity is 2,176 Ci/mmol (80.51 PBq/mol).

A good example of the difference in energy of the various radionuclei is the detection window rnages used to detect them, which are generally proportional to the energy of the emission, but vary from machine to machine: in a Perkin elmer TriLux Beta scintillation counter, the H-3 energy range window is between channel 5–360; C-14, S-35 and P-33 are in the window of 361–660; and P-32 is in the window of 661–1024.

Detection

Quantitative:

- In a liquid scintillation assay (LSA), or liquid scintillation counting (LSC), a small aliquot, filter or swab is added to scintillation fluid and the plate or vial counter in a scintillation counter.

- A Geiger counter is a quick and rough approximation of activity. Lower energy emitters such as tritium can not be detected.

Qualitative

- Autoradiography: A membrane such as a Northern blot or a hybridised slot blot is put against a film that is then developed.

- Phosphor storage screen: The membrane is placed against a phosphor storage screen which is then scanned in a phosphorimager. This is ten times faster and more precise than film and the result is already in digital form.

Microscopy

- Electron microscopy: The sample is not exposed to a beam of electrons but detectors picks up the expelled electrons from the radionuclei.

- Micro-autoradiography imager: A slide is put against scintillation paper and in a PMT. When two different radiolabels are used, a computer can be used to discriminate the two.

Scientific Methods

- Schild regression is a radioligand binding assay. It is used for DNA labelling (5' and 3'), leaving the nucleic acids intact.

Radioactivity Concentration

A vial of radiolabel has a "total activity". Taking as an example γ32P ATP, from the catalogues of the two major suppliers, Perkin Elmer NEG502H500UC or GE AA0068-500UCI, in this case, the total activity is 500 μCi (other typical numbers are 250 μCi or 1 mCi). This is contained in a certain volume, depending on the radioactive concentration, such as 5 to 10 mCi/mL (185 to 370 TBq/m^3); typical volumes include 50 or 25 μL.

Not all molecules in the solution have a P-32 on the last (i.e., gamma) phosphate: the "specific activity" gives the radioactivity concentration and depends on the radionuclei's half-life. If every molecule were labelled, the maximum theoretical specific activity is obtained that for P-32 is 9131 Ci/mmol. Due to pre-calibration and efficiency issues this number is never seen on a label; the values often found are 800, 3000 and 6000 Ci/mmol. With this number it is possible to calculate the total chemical concentration and the hot-to-cold ratio.

"Calibration date" is the date in which the vial's activity is the same as on the label. "Pre-calibration" is the when the activity is calibrated in a future date to compensate for the decay occurred during shipping.

Comparison with Fluorescence

Prior to the widespread use of fluorescence in the past three decades radioactivity was the most common label.

Advantages are:

- fluorescence is much safer and more convenient to use
- Several fluorescent molecules can be used simultaneously (given that they do not overlap, cf. FRET), whereas with radioactivity two isotopes can be used (tritium and a low energy isotope, e.g. ^{33}P due to different intensities) but require special machinery (a tritium screen and a regular phosphor-imaging screen or a specific dual channel detector, e.g.).
- Several properties are extremely useful (cf. next section)

Note: a channel is similar to "colour" but distinct, it is the pair of excitation and emission filters specific for a dye, e.g. agilent microarrays are dual channel, working on cy3 and cy5, these are colloquially referred to as green and red.

Disadvantages are:

- the dye may be a hindrance or toxic.

Safety

If good health physics controls are maintained in a laboratory where radionuclides are used, it is unlikely that the overall radiation dose received by workers will be of much significance. Nevertheless the effects of low doses are mostly unknown so many regulations exist to avoid unnecessary risks, such as skin or internal exposure. Due to the low penetration power and many variables involved it is hard to convert a radioactive concentration to a dose. 1 μCi of P-32 on a square centimetre of skin (through a dead layer of a thickness of 70 μm) gives 7961 rads (79.61 grays) per hour. Similarly a mammogram gives an exposure of 300 mrem (3 mSv) on a larger volume (in the US, the average annual dose is 360 mrem or 3.6 mSv).

Lab-on-a-chip

A lab-on-a-chip (LOC) is a device that integrates one or several laboratory functions on a single chip of only millimeters to a few square centimetres in size. LOCs deal with the handling of extremely small fluid volumes down to less than pico litres. Lab-on-a-chip devices are a subset of MEMS devices and often indicated by "Micro Total Analysis Systems" (µTAS) as well. Microfluidics is a broader term that describes also mechanical flow control devices like pumps and valves or sensors like flowmeters and viscometers. However, strictly regarded "Lab-on-a-Chip" indicates generally the scaling of single or multiple lab processes down to chip-format, whereas "µTAS" is dedicated to the integration of the total sequence of lab processes to perform chemical analysis. The term "Lab-on-a-Chip" was introduced later on when it turned out that µTAS technologies were more widely applicable than only for analysis purposes.

History

After the invention of microtechnology (~1954) for realizing integrated semiconductor structures for microelectronic chips, these lithography-based technologies were soon applied in pressure sensor manufacturing (1966) as well. Due to further development of these usually CMOS-compatibility limited processes, a tool box became available to create micrometre or sub-micrometre sized mechanical structures in silicon wafers as well: the Micro Electro Mechanical Systems (MEMS) era (also indicated with Micro System Technology - MST) had started.

Next to pressure sensors, airbag sensors and other mechanically movable structures, fluid handling devices were developed. Examples are: channels (capillary connections), mixers, valves, pumps and dosing devices. The first LOC analysis system was a gas chromatograph, developed in 1975 by S.C. Terry - Stanford University.

However, only at the end of the 1980s, and beginning of the 1990s, the LOC research started to seriously grow as a few research groups in Europe developed micropumps, flowsensors and the concepts for integrated fluid treatments for analysis systems. These µTAS concepts demonstrated that integration of pre-treatment steps, usually done at lab-scale, could extend the simple sensor functionality towards a complete laboratory analysis, including e.g. additional cleaning and separation steps.

A big boost in research and commercial interest came in the mid 1990's, when µTAS technologies turned out to provide interesting tooling for genomics applications, like capillary electrophoresis and DNA microarrays. A big boost in research support also came from the military, especially from DARPA (Defence Advanced Research Projects Agency), for their interest in portable bio/chemical warfare agent detection systems. The added value was not only limited to integration of lab processes for analysis but also the characteristic possibilities of individual components and the application to other, non-analysis, lab processes. Hence the term "Lab-on-a-Chip" was introduced.

Although the application of LOCs is still novel and modest, a growing interest of companies and applied research groups is observed in different fields such as analysis (e.g. chemical analysis, environmental monitoring, medical diagnostics and cellomics) but also in synthetic chemistry (e.g. rapid screening and microreactors for pharmaceutics). Besides further application developments, research in LOC systems is expected to extend towards downscaling of fluid handling structures as well, by using nanotechnology. Sub-micrometre and nano-sized channels, DNA labyrinths, single cell detection and analysis, and nano-sensors, might become feasible, allowing new ways of interaction with biological species and large molecules.

Chip Materials and Fabrication Technologies

The basis for most LOC fabrication processes is photolithography. Initially most processes were in silicon, as these well-developed technologies were directly derived from semiconductor fabrication. Because of demands for e.g. specific optical characteristics, bio- or chemical compatibility, lower production costs and faster prototyping, new processes have been developed such as glass, ceramics and metal etching, deposition and bonding, PDMS processing (e.g., soft lithography), thick-film- and stereolithography as well as fast replication methods via electroplating, injection molding and embossing. Furthermore the LOC field more and more exceeds the borders between lithography-based microsystem technology, nano technology and precision engineering.it is very useful.

Advantages of LOCs

LOCs may provide advantages, which are specific to their application. Typical advantages are:

- low fluid volumes consumption (less waste, lower reagents costs and less required sample volumes for diagnostics)
- faster analysis and response times due to short diffusion distances, fast heating, high surface to volume ratios, small heat capacities.
- better process control because of a faster response of the system (e.g. thermal control for exothermic chemical reactions)
- compactness of the systems due to integration of much functionality and small volumes
- massive parallelization due to compactness, which allows high-throughput analysis
- lower fabrication costs, allowing cost-effective disposable chips, fabricated in mass production
- safer platform for chemical, radioactive or biological studies because of integration of functionality, smaller fluid volumes and stored energies

Disadvantages of LOCs

- novel technology and therefore not yet fully developed
- physical and chemical effects—like capillary forces, surface roughness, chemical interactions of construction materials on reaction processes—become more dominant on small-scale. This can sometimes make processes in LOCs more complex than in conventional lab equipment
- detection principles may not always scale down in a positive way, leading to low signal-to-noise ratios
- although the absolute geometric accuracies and precision in microfabrication are high, they are often rather poor in a relative way, compared to precision engineering for instance.

Examples of LOC Applications

- Real-time PCR: detect bacteria, viruses and cancers.
- Biochemical assays
- Immunoassay: detect bacteria, viruses and cancers based on antigen-antibody reactions.
- Dielectrophoresis: detection of cancer cells and bacteria.
- Blood sample preparation: can crack cells to extract DNA.
- Cellular lab-on-a-chip for single-cell analysis.

- Ion channel screening (patch clamp)
- Testing the safety and efficacy of new drugs, as with lung on a chip

LOCs and Global Health

Lab-on-a-chip technology may soon become an important part of efforts to improve global health, particularly through the development of point-of-care testing devices. In countries with few healthcare resources, infectious diseases that would be treatable in a developed nation are often deadly. In some cases, poor healthcare clinics have the drugs to treat a certain illness but lack the diagnostic tools to identify patients who should receive the drugs. Many researchers believe that LOC technology may be the key to powerful new diagnostic instruments. The goal of these researchers is to create microfluidic chips that will allow healthcare providers in poorly equipped clinics to perform diagnostic tests such as immunoassays and nucleic acid assays with no laboratory support.

Global Challenges

For the chips to be used in areas with limited resources, many challenges must be overcome. In developed nations, the most highly valued traits for diagnostic tools include speed, sensitivity, and specificity; but in countries where the healthcare infrastructure is less well developed, attributes such ease of use and shelf life must also be considered. The reagents that come with the chip, for example, must be designed so that they remain effective for months even if the chip is not kept in a climate-controlled environment. Chip designers must also keep cost, scalability, and recyclability in mind as they choose what materials and fabrication techniques to use.

Examples of Global LOC Application

One active area of LOC research involves ways to diagnose and manage HIV infections. Around 40 million people are infected with HIV in the world today, yet only 1.3 million of these people receive anti-retroviral treatment. Around 90% of people with HIV have never been tested for the disease. Measuring the number of CD4+ T lymphocytes in a person's blood is an accurate way to determine if a person has HIV and to track the progress of an HIV infection. At the moment, flow cytometry is the gold standard for obtaining CD4 counts, but flow cytometry is a complicated technique that is not available in most developing areas because it requires trained technicians and expensive equipment.

Nuclear Run-on Assay

A nuclear run-on assay is conducted to identify the genes that are being transcribed at a certain time. Cell nuclei are isolated rapidly, and incubated with labelled nucleotides and the results are hybridized to a slot blot, which is then exposed to film. It was originally developed by Gariglio et al. (1981) and Brown et al. (1984). Adding actinomycin-D to the reaction buffer is sufficient to stop transcription, while other toxins, such as α-amanitin, show different effects, making it a good way to assay the effects of these compounds.

This method allows changes in transcription rates to be measured, which often differ from steady-state mRNA levels used in microarrays. Although the method is time-consuming and requires the use of radioactivity and large numbers of cells, it remains the most reliable method to measure transcription rates directly. Several attempts recently have been made to make this method microarray-compatible, but other methods are much more favourable. To have a publication quality blot about 10^7 nuclei are needed, but 10^6 nuclei are sufficient to give results: these quantities of cells are higher than for most protocols.

It is often cited that this method has been surpassed by microarrays, even though this method compares transcription rates and not total transcript expression levels, which differ. It is incompatible with microarray experiments due to the fact eukaryotic polymerases are more intolerant towards fluorescently labelled or aminoallyl nucleotides, so UTP with phosphorus-32, or phosphorus-33, on the alpha phosphate group is normally used. Cheadle et al. (2005) used spotted nylon blots to obtain a high-throughput analysis of T-cell activation. Garcia-Martinez et al (2004)developed a protocol for the yeast S. cerevisiae (Genomic run-on, GRO) that allows for the calculation of transcription rates (TRs) for all yeast genes. Those authors also used the calculated TRs and the steady-state mRNA amounts of all genes to determine, at global scale, the mRNA stabilities for all yeast mRNAs.

Alternative microarray methods have recently been developed, mainly PolII RIP-chip: RNA immunoprecipitation of RNA polymerase II with phosphorylated C-terminal domain directed antibodies and hybridization on a microarray slide or chip (the word chip in the name stems from "ChIP-chip" where a special affymetrix genechip was required). A comparison of methods based on run-on and ChIP-chip has been made in yeast (Pelechano et al., 2009).

A general correspondence of both methods has been detected but GRO is more sensitive and quantitative. It has to be considered that run-on only detects elongating RNA polymerases whereas ChIP-chip detects all present RNA polymerases, including backtracked ones. The main issue on the validity of run on assay results is the presence of artifacts due to the isolation step: some transcripts may stop prematurely others might start out of place. Another point to add is about the nuclei: There are various methods that require nuclear isolation.

The problem is that the nuclear membrane is contiguous with the ER which is rich in protein and cytoplasmic mRNA; some methods require pure nuclear lysate while others, like a nuclear run-on, require "dirty" nuclei with an intact nuclear membrane even if the ER is attached. Some types of cell are notoriously difficult to isolate this way such as lymphocytes, which require a sucrose gradient purification. Yeast cells do not require nuclei isolation. Permeabilized cells (with sarkosil) can be used. This is a clear technical advantage.

Fluorescent in Situ Hybridization

FISH (fluorescence *in situ* hybridization) is a cytogenetic technique developed by Christoph Lengauer that is used to detect and localize the presence or absence of specific DNA sequences on chromosomes. FISH uses fluorescent probes that bind to only those parts of the chromosome with which they show a high degree of sequence similarity. Fluorescence microscopy can be used to find out where the fluorescent probe bound to the chromosomes. FISH is often used for finding specific features in DNA for use in genetic counselling, medicine, and species identification. FISH can also be used to detect and localize specific mRNAs within tissue samples. In this context, it can help define the spatial-temporal patterns of gene expression within cells and tissues.

Probes

Probes are often derived from fragments of DNA that were isolated, purified, and amplified for use in the Human Genome Project. The size of the human genome is so large, compared to the length that could be sequenced directly, that it was necessary to divide the genome into fragments. (In the eventual analysis, these fragments were put into order by digesting a copy of each fragment into still smaller fragments using sequence-specific endonucleases, measuring the size of each small fragment using size-exclusion chromatography, and using that information to determine where the large fragments overlapped one another.) To preserve the fragments with their individual DNA sequences, the fragments were added into a system of continually replicating bacteria populations. Clonal populations of bacteria, each population maintaining a single artificial chromosome, are stored in various laboratories around the world. The artificial chromosomes (BAC) can be grown, extracted, and labelled, in any lab. These fragments are on the order of 100 thousand base-pairs, and are the basis for most FISH probes.

Preparation and Hybridization Process

First, a probe is constructed. The probe must be large enough to hybridize specifically with its target but not so large as to impede the hybridization process. The probe is tagged directly with fluorophores, with targets for antibodies or with biotin. Tagging can be done in various ways, such as nick translation, or PCR using tagged nucleotides.

Then, an interphase or metaphase chromosome preparation is produced. The chromosomes are firmly attached to a substrate, usually glass. Repetitive DNA sequences must be blocked by adding short fragments of DNA to the sample. The probe is then applied to the chromosome DNA and incubated for approximately 12 hours while hybridizing. Several wash steps remove all unhybridized or partially-hybridized probes. The results are then visualized and quantified using a microscope that is capable of exciting the dye and recording images.

If the fluorescent signal is weak, amplification of the signal may be necessary in order to exceed the detection threshold of the microscope. Fluorescent signal strength depends on many factors such as probe labelling efficiency, the type of probe, and the type of dye. Fluorescently-tagged antibodies or streptavidin are bound to the dye molecule. These secondary components are selected so that they have a strong signal.

FISH experiments designed to detect or localize gene expression within cells and tissues rely on the use of a reporter gene, such as one expressing green fluorescent protein, to provide the fluorescence signal.

Variations on Probes and Analysis

FISH is a very general technique. The differences between the various FISH techniques are usually due to variations in the sequence and labelling of the probes; and how they are used in combination. These few modifications make possible all FISH techniques.

Probe size is important because longer probes hybridize more specifically than shorter probes. The overlap defines the resolution of detectable features. For example, if the goal of an experiment is to detect the breakpoint of a translocation, then the overlap of the probes — the degree to which one DNA sequence is contained in the adjacent probes — defines the minimum window in which the breakpoint may be detected.

The mixture of probe sequences determines the type of feature the probe can detect. Probes that hybridize along an entire chromosome are used to count the number of a certain chromosome, show translocations, or identify extra-chromosomal fragments of chromatin. This is often called "whole-chromosome painting." If every possible probe is used, every chromosome, (the whole genome) would be marked fluorescently, which would not be particularly useful for determining features of individual sequences. However, a mixture of smaller probes can be created that is specific to a particular region (locus) of DNA; these mixtures are used to detect deletion mutations. When combined with a specific colour, a locus-specific probe mixture is used to detect very specific translocations. Special locus-specific probe mixtures are often used to count chromosomes, by binding to the centromeric regions of chromosomes, which are unique enough to identify each chromosome (with the exception of Chromosome 13, 14 21, 22.)

A variety of other techniques use mixtures of differently-coloured probes. A range of colors in mixtures of fluorescent dyes can be detected, so each human chromosome can be identified by a characteristic colour using whole-chromosome probe mixtures and a variety of ratios of colors. Although there are more chromosomes than easily-distinguishable fluorescent dye colors, ratios of probe mixtures can be used to create *secondary* colors.

Similar to comparative genomic hybridization, the probe mixture for the secondary colours is created by mixing the correct ratio of two sets of differently-coloured probes for the same chromosome. This technique is sometimes called M-FISH. The same physics that make a variety of colours possible for M-FISH can be used for the detection of translocations. That is, colours that are adjacent appear to overlap; a secondary colour is observed. Some assays are designed so that the secondary colour will be present or absent in cases of interest. An example is the detection of BCR/ABL translocations, where the secondary colour indicates disease. This variation is often called double-fusion FISH or D-FISH. In the opposite situation—where the absence of the secondary colour is pathological—is illustrated by an assay used to investigate translocations where only one of the breakpoints is known or constant. Locus-specific probes are made for one side of the breakpoint and the other intact chromosome. In normal cells, the secondary colour is observed, but only the primary colour is observed when the translocation occurs. This technique is sometimes called "break-apart FISH".

Single Molecule RNA FISH

Single Molecule RNA FISH is a method of detecting and quantifying mRNA and other long RNA molecules in a thin layer of tissue sample. Targets can be reliably imaged through the application of multiple short singly labelled oligonucleotide probes. The probes cooperatively bind to the target site. When each probe binds to the single stranded mRNA, it causes cooperative unwinding of the mRNA, promoting the binding of the next probe. The net result is the binding of 48 fluorescent labels to a single molecule of mRNA, providing sufficient fluorescence to reliably locate each target mRNA in a wide-field fluorescent microscopy image. Probes not binding to the intended sequence do not achieve sufficient localized fluorescence to be distinguished from the background. This technology is exclusively licensed to Biosearch Technologies as Stellaris™ FISH Probes.

Single molecule RNA FISH assays can be performed in simplex or multiplex, and have potential applications in cancer diagnosis, neuroscience, gene expression analysis, and companion diagnostics.

Fiber FISH

In an alternative technique to interphase or metaphase preparations, fiber FISH, interphase chromosomes are attached to a slide in such a way that they are stretched out in a straight line, rather than being tightly coiled, as in conventional FISH, or adopting a random conformation, as in interphase FISH. This is accomplished by applying mechanical shear along the length of the slide, either to cells that have been fixed to the slide and then lysed, or to a solution of purified DNA. A technique known as chromosome combing is increasingly used for this purpose. The extended conformation of the chromosomes allows dramatically higher resolution - even down to a few kilobases. The preparation of fiber FISH samples, although conceptually simple, is a rather skilled art, and only specialized laboratories use the technique routinely.

Q-FISH

Q-FISH combines FISH with PNAs and computer software to quantify fluorescence intensity. This technique is used routinely in telomere length research.

Flow-FISH

Flow-FISH uses flow cytometry to perform FISH automatically using per-cell fluorescence measurements.

Medical Applications

Often parents of children with a developmental delay want to know more about their child's conditions before choosing to have another child. These concerns can be addressed by analysis of the parents' and child's DNA. In cases where the child's developmental delay is not understood, the cause of it can be determined using FISH and cytogenetic techniques.

Examples of diseases that are diagnosed using FISH include Prader-Willi syndrome, Angelman syndrome, 22q13 deletion syndrome, chronic myelogenous leukemia, acute lymphoblastic leukemia, Cri-du-chat, Velocardiofacial syndrome, and Down syndrome. FISH on sperm cells is indicated for men with an abnormal somatic or meiotic karyotype as well as those with oligozoospermia, since approximately 50% of oligozoospermic men have an increased rate of sperm chromosome abnormalities. The analysis of chromosomes 21, X, and Y is enough to identify oligozoospermic individuals at risk.

In medicine, FISH can be used to form a diagnosis, to evaluate prognosis, or to evaluate remission of a disease, such as cancer. Treatment can then be specifically tailored. A traditional exam involving metaphase chromosome analysis is often unable to identify features that distinguish one disease from another, due to subtle chromosomal features; FISH can elucidate these differences.

FISH can also be used to detect diseased cells more easily than standard Cytogenetic methods, which require dividing cells and requires labour and time-intensive manual preparation and analysis of the slides by a technologist. FISH, on the other hand, does not require living cells and can be quantified automatically, a computer counts the fluorescent dots present. However, a trained technologist is required to distinguish subtle differences in banding patterns on bent and twisted metaphase chromosomes.

Species Identification

FISH is often used in clinical studies. If a patient is infected with a suspected pathogen, bacteria, from the patient's tissues or fluids, are typically grown on agar to determine the identity of the pathogen. Many bacteria, however, even well-known species, do not grow well under laboratory conditions. FISH can be used to detect directly the presence of the suspect on small samples of patient's tissue.

FISH can also be used to compare the genomes of two biological species, to deduce evolutionary relationships. A similar hybridization technique is called a zoo blot. Bacterial FISH probes are often primers for the 16s rRNA region.

FISH is widely used in the field of microbial ecology, to identify microorganisms. Biofilms, for example, are composed of complex (often) multi-species bacterial organizations. Preparing DNA probes for one species and performing FISH with this probe allows one to visualize the distribution of this specific species within the biofilm. Preparing probes (in two different colours) for two species allows to visualize/study co-localization of these two species in the biofilm, and can be useful in determining the fine architecture of the biofilm.

Lab-on-a-chip and FISH

Although interphase fluorescence in situ hybridization (FISH) is a sensitive diagnostic tool used for the detection of chromosomal abnormalities on cell-by-cell basis, the cost-per-test and the technical complexity of current FISH protocols has inhibited its widespread utilization. Lab-on-a-chip or microfluidic devices, incorporate networks of microchannels that can miniaturize, integrate and automate conventional analytical techniques onto chip-style platforms. Since microchannels permit sophisticated levels of fluid control (down to picolitres), these devices can reduce analysis times, lower reagent consumption, and minimize human intervention.

Currently, FISH has been performed on glass microfluidic platforms that standardize much of the protocol offering repeatable results that are accurate, cost-effective and easier to obtain in a clinical setting. Compared to conventional FISH methods, these first implementations of on-chip FISH provide a 10-fold higher throughput and a 10-fold reduction in the cost of testing, enabling the simultaneous assessment of several chromosomal abnormalities or patients. It is increasingly essential that diagnostic tests determine the type and extent of chromosomal abnormalities for more informed diagnosis and for appropriate choice of treatment strategies. Since the on-chip FISH technique is 10-20 times more cost-effective than conventional methods, and can be fully integrated and automated, this technology will make widespread genetic testing of patients more accessible in a clinical setting.

Virtual Karyotype

Virtual karyotyping is another cost-effective, clinically available alternative to FISH panels uses thousands to millions of probes on a single array to detect copy number changes, genome-wide, at unprecedented resolution.

Bioinformatics

Bioinformatics is the application of statistics and computer science to the field of molecular biology.

The term *bioinformatics* was coined by Paulien Hogeweg and Ben Hesper in 1978 for the study of informatic processes in biotic systems. Its primary use since at least the late 1980s has been in genomics and genetics, particularly in those areas of genomics involving large-scale DNA sequencing.

Bioinformatics now entails the creation and advancement of databases, algorithms, computational and statistical techniques and theory to solve formal and practical problems arising from the management and analysis of biological data.

Over the past few decades rapid developments in genomic and other molecular research technologies and developments in information technologies have combined to produce a tremendous amount of information related to molecular biology. It is the name given to these mathematical and computing approaches used to glean understanding of biological processes.

Common activities in bioinformatics include mapping and analysing DNA and protein sequences, aligning different DNA and protein sequences to compare them and creating and viewing 3-D models of protein structures. The primary goal of bioinformatics is to increase the understanding of biological processes. What sets it apart from other approaches, however, is its focus on developing and applying computationally intensive techniques (e.g., pattern recognition, data mining, machine learning algorithms, and visualization) to achieve this goal. Major research efforts in the field include sequence alignment, gene finding, genome assembly, drug design, drug discovery, protein structure alignment, protein structure prediction, prediction of gene expression and protein-protein interactions, genome-wide association studies and the modeling of evolution.

Introduction

Bioinformatics was applied in the creation and maintenance of a database to store biological information at the beginning of the "genomic revolution", such as nucleotide and amino acid sequences. Development of this type of database involved not only design issues but the development of complex interfaces whereby researchers could both access existing data as well as submit new or revised data.

In order to study how normal cellular activities are altered in different disease states, the biological data must be combined to form a comprehensive picture of these activities. Therefore, the field of bioinformatics has evolved such that the most pressing task now involves the analysis and interpretation of various types of data, including nucleotide and amino acid sequences, protein domains, and protein structures.

The actual process of analysing and interpreting data is referred to as computational biology. Important sub-disciplines within bioinformatics and computational biology include:

- the development and implementation of tools that enable efficient access to, and use and management of, various types of information.
- the development of new algorithms (mathematical formulas) and statistics with which to assess relationships among members of large data sets, such as methods to locate a gene within a sequence, predict protein structure and/or function, and cluster protein sequences into families of related sequences.

There are two fundamental ways of modelling a Biological system (e.g. living cell) both coming under Bioinformatic approaches.

- Static
  - Sequences - Proteins, Nucleic acids and Peptides
  - Structures - Proteins, Nucleic acids, Ligands (including metabolites and drugs) and Peptides
  - Interaction data among the above entities including microarray data and Networks of proteins, metabolites

- Dynamic
  - o Systems Biology comes under this category including reaction fluxes and variable concentrations of metabolites
  - o Multi-Agent Based modelling approaches capturing cellular events such as signalling, transcription and reaction dynamics.

A broad sub-category under bioinformatics is structural bioinformatics.

Major Research Areas

Sequence Analysis

Since the Phage Φ-X174 was sequenced in 1977, the DNA sequences of thousands of organisms have been decoded and stored in databases. This sequence information is analysed to determine genes that encode polypeptides (proteins), RNA genes, regulatory sequences, structural motifs, and repetitive sequences. A comparison of genes within a species or between different species can show similarities between protein functions, or relations between species (the use of molecular systematics to construct phylogenetic trees). With the growing amount of data, it long ago became impractical to analyze DNA sequences manually. Today, computer programs such as BLAST are used daily to search sequences from more than 260 000 of organisms, containing over 190 billion nucleotides. These programs can compensate for mutations (exchanged, deleted or inserted bases) in the DNA sequence, in order to identify sequences that are related, but not identical. A variant of this sequence alignment is used in the sequencing process itself. The so-called shotgun sequencing technique (which was used, for example, by The Institute for Genomic Research to sequence the first bacterial genome, *Haemophilus influenzae*) does not produce entire chromosomes, but instead generates the sequences of many thousands of small DNA fragments (ranging from 35 to 900 nucleotides long, depending on the sequencing technology). The ends of these fragments overlap and, when aligned properly by a genome assembly program, can be used to reconstruct the complete genome. Shotgun sequencing yields sequence data quickly, but the task of assembling the fragments can be quite complicated for larger genomes. For a genome as large as the human genome, it may take many days of CPU time on large-memory, multiprocessor computers to assemble the fragments, and the resulting assembly will usually contain numerous gaps that have to be filled in later. Shotgun sequencing is the method of choice for virtually all genomes sequenced today, and genome assembly algorithms are a critical area of bioinformatics research.

Another aspect of bioinformatics in sequence analysis is annotation, which involves computational gene finding to search for protein-coding genes, RNA genes, and other functional sequences within a genome. Not all of the nucleotides within a genome are part of genes. Within the genome of higher organisms, large parts of the DNA do not serve any obvious purpose. This so-called junk DNA may, however, contain unrecognized functional elements. Bioinformatics helps to bridge the gap between genome and proteome projects — for example, in the use of DNA sequences for protein identification.

Genome Annotation

In the context of genomics, annotation is the process of marking the genes and other biological features in a DNA sequence. The first genome annotation software system was designed in 1995 by Dr. Owen White, who was part of the team at The Institute for Genomic Research that sequenced and analysed the first genome of a free-living organism to be decoded, the bacterium *Haemophilus influenzae*. Dr. White built a software system to find the genes (places in the DNA sequence that encode a protein), the transfer RNA, and other features, and to make initial assignments of function to those genes. Most current genome annotation systems work similarly, but the programs available for analysis of genomic DNA are constantly changing and improving.

Computational Evolutionary Biology

Evolutionary biology is the study of the origin and descent of species, as well as their change over time. Informatics has assisted evolutionary biologists in several key ways; it has enabled researchers to:

- trace the evolution of a large number of organisms by measuring changes in their DNA, rather than through physical taxonomy or physiological observations alone,

- more recently, compare entire genomes, which permits the study of more complex evolutionary events, such as gene duplication, horizontal gene transfer, and the prediction of factors important in bacterial speciation,

- build complex computational models of populations to predict the outcome of the system over time
- track and share information on an increasingly large number of species and organisms

Future work endeavours to reconstruct the now more complex tree of life.

The area of research within computer science that uses genetic algorithms is sometimes confused with computational evolutionary biology, but the two areas are not necessarily related.

Analysis of Gene Expression

The expression of many genes can be determined by measuring mRNA levels with multiple techniques including microarrays, expressed cDNA sequence tag (EST) sequencing, serial analysis of gene expression (SAGE) tag sequencing, massively parallel signature sequencing (MPSS), or various applications of multiplexed in-situ hybridization. All of these techniques are extremely noise-prone and/ or subject to bias in the biological measurement, and a major research area in computational biology involves developing statistical tools to separate signal from noise in high-throughput gene expression studies. Such studies are often used to determine the genes implicated in a disorder: one might compare microarray data from cancerous epithelial cells to data from non-cancerous cells to determine the transcripts that are up-regulated and down-regulated in a particular population of cancer cells.

Analysis of Regulation

Regulation is the complex orchestration of events starting with an extracellular signal such as a hormone and leading to an increase or decrease in the activity of one or more proteins. Bioinformatics techniques have been applied to explore various steps in this process. For example, promoter analysis involves the identification and study of sequence motifs in the DNA surrounding the coding region of a gene. These motifs influence the extent to which that region is transcribed into mRNA. Expression data can be used to infer gene regulation: one might compare microarray data from a wide variety of states of an organism to form hypotheses about the genes involved in each state. In a single-cell organism, one might compare stages of the cell cycle, along with various stress conditions (heat shock, starvation, etc.). One can then apply clustering algorithms to that expression data to determine which genes are co-expressed. For example, the upstream regions (promoters) of co-expressed genes can be searched for over-represented regulatory elements.

Analysis of Protein Expression

Protein microarrays and high throughput (HT) mass spectrometry (MS) can provide a snapshot of the proteins present in a biological sample. Bioinformatics is very much involved in making sense of protein microarray and HT MS data; the former approach faces similar problems as with microarrays targeted at mRNA, the latter involves the problem of matching large amounts of mass data against predicted masses from protein sequence databases, and the complicated statistical analysis of samples where multiple, but incomplete peptides from each protein are detected.

Analysis of Mutations in Cancer

In cancer, the genomes of affected cells are rearranged in complex or even unpredictable ways. Massive sequencing efforts are used to identify previously unknown point mutations in a variety of genes in cancer. Bioinformaticians continue to produce specialized automated systems to manage the sheer volume of sequence data produced, and they create new algorithms and software to compare the sequencing results to the growing collection of human genome sequences and germline polymorphisms. New physical detection technologies are employed, such as oligonucleotide microarrays to identify chromosomal gains and losses (called comparative genomic hybridization), and single-nucleotide polymorphism arrays to detect known *point mutations*. These detection methods simultaneously measure several hundred thousand sites throughout the genome, and when used in high-throughput to measure thousands of samples, generate terabytes of data per experiment. Again the massive amounts and new types of data generate new opportunities for bioinformaticians. The data is often found to contain considerable variability, or noise, and thus Hidden Markov model and change-point analysis methods are being developed to infer real copy number changes.

Another type of data that requires novel informatics development is the analysis of lesions found to be recurrent among many tumours.

Comparative Genomics

The core of comparative genome analysis is the establishment of the correspondence between genes (orthology analysis) or other genomic features in different organisms. It is these intergenomic maps that make it possible to trace the evolutionary processes responsible for the divergence of two genomes. A multitude of evolutionary events acting at various organizational levels shape genome evolution. At the lowest level, point mutations affect individual nucleotides. At a higher level, large chromosomal segments undergo duplication, lateral transfer, inversion, transposition, deletion and insertion. Ultimately, whole genomes are involved in processes of hybridization, polyploidization and endosymbiosis, often leading to rapid speciation. The complexity of genome evolution poses many exciting challenges to developers of mathematical models and algorithms, who have recourse to a spectra of algorithmic, statistical and mathematical techniques, ranging from exact, heuristics, fixed parameter and approximation algorithms for problems based on parsimony models to Markov Chain Monte Carlo algorithms for Bayesian analysis of problems based on probabilistic models.

Many of these studies are based on the homology detection and protein families computation.

Modeling Biological Systems

Systems biology involves the use of computer simulations of cellular subsystems (such as the networks of metabolites and enzymes which comprise metabolism, signal transduction pathways and gene regulatory networks) to both analyze and visualize the complex connections of these cellular processes. Artificial life or virtual evolution attempts to understand evolutionary processes via the computer simulation of simple (artificial) life forms.

High-throughput Image Analysis

Computational technologies are used to accelerate or fully automate the processing, quantification and analysis of large amounts of high-information-content biomedical imagery. Modern image analysis systems augment an observer's ability to make measurements from a large or complex set of images, by improving accuracy, objectivity, or speed. A fully developed analysis system may completely replace the observer. Although these systems are not unique to biomedical imagery, biomedical imaging is becoming more important for both diagnostics and research. Some examples are:

- high-throughput and high-fidelity quantification and sub-cellular localization (high-content screening, cytohistopathology, Bioimage informatics)

- morphometrics
- clinical image analysis and visualization
- determining the real-time air-flow patterns in breathing lungs of living animals
- quantifying occlusion size in real-time imagery from the development of and recovery during arterial injury
- making behavioural observations from extended video recordings of laboratory animals
- infrared measurements for metabolic activity determination
- inferring clone overlaps in DNA mapping, e.g. the Sulston score.

Structural Bioinformatic Approaches

Prediction of Protein Structure

Protein structure prediction is another important application of bioinformatics. The amino acid sequence of a protein, the so-called primary structure, can be easily determined from the sequence on the gene that codes for it. In the vast majority of cases, this primary structure uniquely determines a structure in its native environment. (Of course, there are exceptions, such as the bovine spongiform encephalopathy - aka Mad Cow Disease - prion.) Knowledge of this structure is vital in understanding the function of the protein. For lack of better terms, structural information is usually classified as one of *secondary*, *tertiary* and *quaternary* structure. A viable general solution to such predictions remains an open problem. As of now, most efforts have been directed towards heuristics that work most of the time.

One of the key ideas in bioinformatics is the notion of homology. In the genomic branch of bioinformatics, homology is used to predict the function of a gene: if the sequence of gene *A*, whose function is known, is homologous to the sequence of gene *B,* whose function is unknown, one could infer that B may share A's function. In the structural branch of bioinformatics, homology is used to determine which parts of a protein are important in structure formation and interaction with other proteins.

In a technique called homology modeling, this information is used to predict the structure of a protein once the structure of a homologous protein is known. This currently remains the only way to predict protein structures reliably.

One example of this is the similar protein homology between hemoglobin in humans and the hemoglobin in legumes (leghemoglobin). Both serve the same purpose of transporting oxygen in the organism. Though both of these proteins have completely different amino acid sequences, their protein structures are virtually identical, which reflects their near identical purposes.

Other techniques for predicting protein structure include protein threading and *de novo* (from scratch) physics-based modeling.

Molecular Interaction

Efficient software is available today for studying interactions among proteins, ligands and peptides. Types of interactions most often encountered in the field include - Protein-ligand (including drug), protein-protein and protein-peptide.

Molecular dynamic simulation of movement of atoms about rotatable bonds is the fundamental principle behind computational algorithms, termed docking algorithms for studying molecular interactions.

Docking Algorithms

In the last two decades, tens of thousands of protein three-dimensional structures have been determined by X-ray crystallography and Protein nuclear magnetic resonance spectroscopy (protein NMR). One central question for the biological scientist is whether it is practical to predict possible protein-protein interactions only based on these 3D shapes, without doing protein-protein interaction experiments. A variety of methods have been developed to tackle the Protein-protein docking problem, though it seems that there is still much work to be done in this field.

Software and Tools

Software tools for bioinformatics range from simple command-line tools, to more complex graphical programs and standalone web-services available from various bioinformatics companies or public institutions.

Web Services in Bioinformatics

SOAP and REST-based interfaces have been developed for a wide variety of bioinformatics applications allowing an application running on one computer in one part of the world to use algorithms, data and computing resources on servers in other parts of the world. The main advantages derive from the fact that end users do not have to deal with software and database maintenance overheads. Basic bioinformatics services are classified by the EBI into three categories: SSS (Sequence Search Services), MSA (Multiple Sequence Alignment) and BSA (Biological Sequence Analysis).

The availability of these service-oriented bioinformatics resources demonstrate the applicability of web based bioinformatics solutions, and range from a collection of standalone tools with a common data format under a single, standalone or web-based interface, to integrative, distributed and extensible bioinformatics workflow management systems.

7

Biochemistry of Nucleic Acid

Nucleic acids are biological molecules essential for life, and include DNA (deoxyribonucleic acid) and RNA (ribonucleic acid). Together with proteins, nucleic acids make up the most important macromolecules; each is found in abundance in all living things. Nucleic acids were first discovered by Friedrich Miescher in 1871. Experimental studies of nucleic acids constitute a major part of modern biological and medical research, and form a foundation for genome and forensic science, as well as the biotechnology and pharmaceutical industries.

Nucleic acid, naturally occurring chemical compound that is capable of being broken down to yield phosphoric acid, sugars, and a mixture of organic bases (purines and pyrimidines). Nucleic acids are the main information-carrying molecules of the cell, and, by directing the process of protein synthesis, they determine the inherited characteristics of every living thing. The two main classes of nucleic acids are deoxyribonucleic acid (DNA) and ribonucleic acid (RNA). DNA is the master blueprint for life and constitutes the genetic material in all free-living organisms and most viruses. RNA is the genetic material of certain viruses, but it is also found in all living cells, where it plays an important role in certain processes such as the making of proteins.

Nucleotides: Building Blocks of Nucleic Acids

Basic Structure

Nucleic acids are polynucleotides—that is, long chainlike molecules composed of a series of nearly identical building blocks called nucleotides. Each nucleotide consists of a nitrogen-containing aromatic base attached to a pentose (five-carbon) sugar, which is in turn attached to a phosphate group. Each nucleic acid contains four of five possible nitrogen-containing bases: adenine (A), guanine (G), cytosine (C), thymine (T), and uracil (U). A and G are categorized as purines, and C, T, and U are collectively called pyrimidines. All nucleic acids contain the bases A, C, and G; T, however, is found only in DNA, while U is found in RNA. The pentose sugar in DNA (22 -deoxyribose) differs from the sugar in RNA (ribose) by the absence of a hydroxyl group ("OH) on the 22 carbon of the sugar ring. Without an attached phosphate group, the sugar attached to one of the bases is known as a nucleoside. The phosphate group connects successive sugar residues by bridging the 52 -hydroxyl group on one sugar to the 32 -hydroxyl group of the next sugar in the chain. These nucleoside linkages are called phosphodiester bonds and are the same in RNA and DNA.

Biosynthesis and Degradation

Nucleotides are synthesized from readily available precursors in the cell. The ribose phosphate portion of both purine and pyrimidine nucleotides is synthesized from glucose via the pentose phosphate pathway. The six-atom pyrimidine ring is synthesized first and subsequently attached to the ribose phosphate. The two rings in purines are synthesized while attached to the ribose phosphate during the assembly of adenine or guanine nucleosides. In both cases the end product is a nucleotide carrying a phosphate attached to the 52 carbon on the sugar. Finally, a specialized enzyme called a kinase adds two phosphate groups using adenosine triphosphate (ATP) as the phosphate donor to form ribonucleoside triphosphate, the immediate precursor of RNA. For DNA, the 22 -hydroxyl group is removed from the ribonucleoside diphosphate to give deoxyribonucleoside diphosphate. An additional phosphate group from ATP is then added by another kinase to form a deoxyribonucleoside triphosphate, the immediate precursor of DNA.

During normal cell metabolism, RNA is constantly being made and broken down. The purine and pyrimidine residues are reused by several salvage pathways to make more genetic material. Purine is salvaged in the form of the corresponding nucleotide, whereas pyrimidine is salvaged as the nucleoside.

Deoxyribonucleic Acid (DNA)

DNA is a polymer of the four nucleotides A, C, G, and T, which are joined through a backbone of alternating phosphate and deoxyribose sugar residues. These nitrogen-containing bases occur in complementary pairs as determined by their ability to form hydrogen bonds between them. A always pairs with T through two hydrogen bonds, and G always pairs with C through three hydrogen bonds. The spans of A:T and G:C hydrogen-bonded pairs are nearly identical, allowing them to bridge the sugar-phosphate chains uniformly. This structure, along with the molecule's chemical stability, makes DNA the ideal genetic material. The bonding between complementary bases also provides a mechanism for the replication of DNA and the transmission of genetic information.

Chemical Structure

In 1953 James D. Watson and Francis H.C. Crick proposed a three-dimensional structure for DNA based on low-resolution X-ray crystallographic data and on Erwin Chargaff's observation that, in naturally occurring DNA, the amount of T equals the amount of A and the amount of G equals the amount of C. Watson and Crick, who shared a Nobel Prize in 1962 for their efforts, postulated that two strands of polynucleotides coil around each other, forming a double helix. The two strands, though identical, run in opposite directions as determined by the orientation of the 52 to 32 phosphodiester bond. The sugar-phosphate chains run along the outside of the helix, and the bases lie on the inside, where they are linked to complementary bases on the other strand through hydrogen bonds.

The double helical structure of normal DNA takes a right-handed form called the B-helix. The helix makes one complete turn approximately every 10 base pairs. B-DNA has two principal grooves, a wide major groove and a narrow minor groove. Many proteins interact in the space of the major groove, where they make sequence-specific contacts with the bases. In addition, a few proteins are known to make contacts via the minor groove.

Several structural variants of DNA are known. In A-DNA, which forms under conditions of high salt concentration and minimal water, the base pairs are tilted and displaced toward the minor groove. Left-handed Z-DNA forms most readily in strands that contain sequences with alternating purines and pyrimidines. DNA can form triple helices when two strands containing runs of pyrimidines interact with a third strand containing a run of purines.

B-DNA is generally depicted as a smooth helix; however, specific sequences of bases can distort the otherwise regular structure. For example, short tracts of A residues interspersed with short sections of general sequence result in a bent DNA molecule. Inverted base sequences, on the other hand, produce cruciform structures with four-way junctions that are similar to recombination intermediates. Most of these alternative DNA structures have only been characterized in the laboratory, and their cellular significance is unknown.

Biological Structures

Naturally occurring DNA molecules can be circular or linear. The genomes of single-celled bacteria and archaea (the prokaryotes), as well as the genomes of mitochondria and chloroplasts (certain functional structures within the cell), are circular molecules. In addition, some bacteria and archaea have smaller circular DNA molecules called plasmids that typically contain only a few genes. Many plasmids are readily transmitted from one cell to another. For a typical bacterium, the genome that encodes all of the genes of the organism is a single contiguous circular molecule that contains a half million to five million base pairs. The genomes of most eukaryotes and some prokaryotes contain linear DNA molecules called chromosomes. Human DNA, for example, consists of 23 pairs of linear chromosomes containing three billion base pairs.

In all cells, DNA does not exist free in solution but rather as a protein-coated complex called chromatin. In prokaryotes, the loose coat of proteins on the DNA helps to shield the negative charge of the phosphodiester backbone. Chromatin also contains proteins that control gene expression and determine the characteristic shapes of chromosomes. In eukaryotes, a section of DNA between 140 and 200 base pairs long winds around a discrete set of eight positively charged proteins called a histone, forming a spherical structure called the nucleosome. Additional histones are wrapped by successive sections of DNA, forming a series of nucleosomes like beads on a string. Transcription and replication of DNA is more complicated in eukaryotes because the nucleosome complexes have to be at least partially disassembled for the processes to proceed effectively.

Most prokaryote viruses contain linear genomes that typically are much shorter and contain only the genes necessary for viral propagation. Bacterial viruses called bacteriophages (or phages) may contain both linear and circular forms of DNA. For instance, the genome of bacteriophage λ (lambda), which infects the bacterium *Escherichia coli*, contains 48,502 base pairs and can exist as a linear molecule packaged in a protein coat. The DNA of phage λ can also exist in a circular form (as described in the section Site-specific recombination) that is able to integrate into the circular genome of the host bacterial cell. Both circular and linear genomes are found among eukaryotic viruses, but they more commonly use RNA as the genetic material.

Biochemical Properties

Denaturation

The strands of the DNA double helix are held together by hydrogen bonding interactions between the complementary base pairs. Heating DNA in solution easily breaks these hydrogen bonds, allowing the two strands to separate—a process called denaturation or melting. The two strands may reassociate when the solution cools, reforming the starting DNA duplex—a process called renaturation or hybridization. These processes form the basis of many important techniques for manipulating DNA. For example, a short piece of DNA called an oligonucleotide can be used to test whether a very long DNA sequence has the complementary sequence of the oligonucleotide embedded within it. Using hybridization, a single-stranded DNA molecule can capture complementary sequences from any source. Single strands from RNA can also reassociate. DNA and RNA single strands can form hybrid molecules that are even more stable than double-stranded DNA. These molecules form the basis of a technique that is used to purify and characterize messenger RNA (mRNA) molecules corresponding to single genes.

Ultraviolet Absorption

DNA melting and reassociation can be monitored by measuring the absorption of ultraviolet (UV) light at a wavelength of 260 nanometres (billionths of a metre). When DNA is in a double-stranded conformation, absorption is fairly weak, but when DNA is single-stranded, the unstacking of the bases leads to an enhancement of absorption called hyperchromicity. Therefore, the extent to which DNA is single-stranded or double-stranded can be determined by monitoring UV absorption.

Chemical Modification

After a DNA molecule has been assembled, it may be chemically modified—sometimes deliberately by special enzymes called DNA methyltransferases and sometimes accidentally by oxidation, ionizing radiation, or the action of chemical carcinogens. DNA can also be cleaved and degraded by enzymes called nucleases.

Methylation

Three types of natural methylation have been reported in DNA. Cytosine can be modified either on the ring to form 5-methylcytosine or on the exocyclic amino group to form N^4-methylcytosine. Adenine may be modified to form N^6-methyladenine. N^4-methylcytosine and N^6-methyladenine are found only in bacteria and archaea, whereas 5-methylcytosine is widely distributed. Special enzymes called DNA methyltransferases are responsible for this methylation; they recognize specific sequences within the DNA molecule so that only a subset of the bases is modified. Other methylations of the bases or of the deoxyribose are sometimes induced by carcinogens. These usually lead to mispairing of the bases during replication and have to be removed if they are not to become mutagenic.

Natural methylation has many cellular functions. In bacteria and archaea, methylation forms an essential part of the immune system by protecting DNA molecules from fragmentation by restriction endonucleases. In some organisms, methylation helps to eliminate incorrect base sequences introduced during DNA replication. By marking the parental strand with a methyl group, a cellular mechanism known as the mismatch repair system distinguishes between the newly replicated strand where the errors occur and the correct sequence on the template strand.

In higher eukaryotes, 5-methylcytosine controls many cellular phenomena by preventing DNA transcription. Methylation is also believed to signal imprinting, a process whereby some genes inherited from one parent are selectively inactivated. Correct methylation may also repress or activate key genes that control embryonic development. On the other hand, 5-methylcytosine is potentially mutagenic because thymine produced during the methylation process converts C:G pairs to T:A pairs. In mammals, methylation takes place selectively within the dinucleotide sequence CG—a rare sequence, presumably because it has been lost by mutation. In many cancers, mutations are found in key genes at CG dinucleotides.

Nucleases

Nucleases are enzymes that hydrolytically cleave the phosphodiester backbone of DNA. Endonucleases cleave in the middle of chains, while exonucleases operate selectively by degrading from the end of the chain. Nucleases that act on both single- and double-stranded DNA are known.

Restriction endonucleases are a special class that recognize and cleave specific sequences in DNA. Type II restriction endonucleases always cleave at or near their recognition sites. They produce small, well-defined fragments of DNA that help to characterize genes and genomes and that produce recombinant DNAs. Fragments of DNA produced by restriction endonucleases can be moved from one organism to another. In this way it has been possible to express proteins such as human insulin in bacteria.

Mutation

Chemical modification of DNA can lead to mutations in the genetic material. Anions such as bisulfite can deaminate cytosine to form uracil, changing the genetic message by causing C-to-T transitions. Exposure to acid causes the loss of purine residues, though specific enzymes exist in cells to repair these lesions. Exposure to UV light can cause adjacent pyrimidines to dimerize, while oxidative damage from free radicals or strong oxidizing agents can cause a variety of lesions that are mutagenic if not repaired. Halogens such as chlorine and bromine react directly with uracil, adenine, and guanine, giving substituted bases that are often mutagenic. Similarly, nitrous acid reacts with primary amine groups—for example, converting adenosine into inosine—which then leads to changes in base pairing and mutation. Many chemical mutagens, such as chlorinated hydrocarbons and nitrites, owe their toxicity to the production of halides and nitrous acid during their metabolism in the body.

Supercoiling

Circular DNA molecules such as those found in plasmids or bacterial chromosomes can adopt many different topologies. One is active supercoiling, which involves the cleavage of one DNA strand, its winding one or more turns around the complementary strand, and then the resealing of the molecule. Each complete rotation leads to the introduction of one supercoiled turn in the DNA, a process that can continue until the DNA is fully wound and collapses on itself in a tight ball. Reversal is also possible. Special enzymes called gyrases and topoisomerases catalyse the winding and relaxation of supercoiled DNA. In the linear chromosomes of eukaryotes, the DNA is usually tightly constrained at various points by proteins, allowing the intervening stretches to be supercoiled. This property is partially responsible for the great compaction of DNA that is necessary to fit it within the confines of the cell. The DNA in one human cell would have an extended length of between two and three metres, but it is packed very tightly so that it can fit within a human cell nucleus that is 10 micrometres in diameter.

Sequence Determination

Methods to determine the sequences of bases in DNA were pioneered in the 1970s by Frederick Sanger and Walter Gilbert, whose efforts won them a Nobel Prize in 1980. The Gilbert-Maxam method relies on the different chemical reactivities of the bases, while the Sanger method is based on enzymatic synthesis of DNA in vitro. Both methods measure the distance from a fixed point on DNA to each occurrence of a particular base—A, C, G, or T.

DNA fragments obtained from a series of reactions are separated according to length in four "lanes" by gel electrophoresis. Each lane corresponds to a unique base, and the sequence is read directly from the gel. The Sanger method has now been automated using fluorescent dyes to label the DNA, and a single machine can produce tens of thousands of DNA base sequences in a single run.

Ribonucleic Acid (RNA)

RNA is a single-stranded nucleic acid polymer of the four nucleotides A, C, G, and U joined through a backbone of alternating phosphate and ribose sugar residues. It is the first intermediate in converting the information from DNA into proteins essential for the working of a cell. Some RNAs also serve direct roles in cellular metabolism. RNA is made by copying the base sequence of a section of double-stranded DNA, called a gene, into a piece of single-stranded nucleic acid. This process, called transcription, is catalysed by an enzyme called RNA polymerase.

Chemical Structure

Whereas DNA provides the genetic information for the cell and is inherently quite stable, RNA has many roles and is much more reactive chemically. RNA is sensitive to oxidizing agents such as periodate that lead to opening of the 32 -terminal ribose ring. The 22 -hydroxyl group on the ribose ring is a major cause of instability in RNA, because the presence of alkali leads to rapid cleavage of the phosphodiester bond linking ribose and phosphate groups. In general, this instability is not a significant problem for the cell, because RNA is constantly being synthesized and degraded.

Interactions between the nitrogen-containing bases differ in DNA and RNA. In DNA, which is usually double-stranded, the bases in one strand pair with complementary bases in a second DNA strand. In RNA, which is usually single-stranded, the bases pair with other bases within the same molecule, leading to complex three-dimensional structures. Occasionally, intermolecular RNA/RNA duplexes do form, but they form a right-handed A-type helix rather than the B-type DNA helix.

Depending on the amount of salt present, either 11 or 12 base pairs are found in each turn of the helix. Helices between RNA and DNA molecules also form; these adopt the A-type conformation and are more stable than either RNA/RNA or DNA/DNA duplexes. Such hybrid duplexes are important species in biology, being formed when RNA polymerase transcribes DNA into mRNA for protein synthesis and when reverse transcriptase copies a viral RNA genome such as that of the human immunodeficiency virus (HIV).

Single-stranded RNAs are flexible molecules that form a variety of structures through internal base pairing and additional non-base pair interactions. They can form hairpin loops such as those found in transfer RNA (tRNA), as well as longer-range interactions involving both the bases and the phosphate residues of two or more nucleotides. This leads to compact three-dimensional structures. Most of these structures have been inferred from biochemical data, since few crystallographic images are available for RNA molecules. In some types of RNA, a large number of bases are modified after the RNA is transcribed. More than 90 different modifications have been documented, including extensive methylations and a wide variety of substitutions around the ring. In some cases these modifications are known to affect structure and are essential for function.

Types of RNA

Messenger RNA (mRNA)

Messenger RNA (mRNA) delivers the information encoded in one or more genes from the DNA to the ribosome, a specialized structure, or organelle, where that information is decoded into a protein. In prokaryotes, mRNAs contain an exact transcribed copy of the original DNA sequence with a terminal 52 -triphosphate group and a 32 - hydroxyl residue. In eukaryotes the mRNA molecules are more elaborate. The 52 -triphosphate residue is further esterified, forming a structure called a cap. At the 32 ends, eukaryotic mRNAs typically contain long runs of adenosine residues (polyA) that are not encoded in the DNA but are added enzymatically after transcription. Eukaryotic mRNA molecules are usually composed of small segments of the original gene and are generated by a process of cleavage and rejoining from an original precursor RNA (pre-mRNA) molecule, which is an exact copy of the gene (as described in the section Splicing). In general, prokaryotic mRNAs are degraded very rapidly, whereas the cap structure and the polyA tail of eukaryotic mRNAs greatly enhance their stability.

Ribosomal RNA (rRNA)

Ribosomal RNA (rRNA) molecules are the structural components of the ribosome. The rRNAs form extensive secondary structures and play an active role in recognizing conserved portions of mRNAs and tRNAs. They also assist with the catalysis of protein synthesis. In the prokaryote *E. coli*, seven copies of the rRNA genes synthesize about 15,000 ribosomes per cell. In eukaryotes the numbers are much larger. Anywhere from 50 to 5,000 sets of rRNA genes and as many as 10 million ribosomes may be present in a single cell. In eukaryotes these rRNA genes are looped out of the main chromosomal fibres and coalesce in the presence of proteins to form an organelle called the nucleolus. The nucleolus is where the rRNA genes are transcribed and the early assembly of ribosomes takes place.

Transfer RNA (tRNA)

Transfer RNA (tRNA) carries individual amino acids into the ribosome for assembly into the growing polypeptide chain. The tRNA molecules contain 70 to 80 nucleotides and fold into a characteristic cloverleaf structure. Specialized tRNAs exist for each of the 20 amino acids needed for protein synthesis, and in many cases more than one tRNA for each amino acid is present. The nucleotide sequence is converted into a protein sequence by translating each three-base sequence (called a codon) with a specific protein. The 61 codons used to code amino acids can be read by many fewer than 61 distinct tRNAs (as described in the section Translation). In *E. coli* a total of 40 different tRNAs are used to translate the 61 codons. The amino acids are loaded onto the tRNAs by specialized enzymes called aminoacyl tRNA synthetases, usually with one synthetase for each amino acid. However, in some organisms, less than the full complement of 20 synthetases are required because some amino acids, such as glutamine and asparagine, can be synthesized on their respective tRNAs. All tRNAs adopt similar structures because they all have to interact with the same sites on the ribosome.

Ribozymes

Not all catalysis within the cell is carried out exclusively by proteins. Thomas Cech and Sidney Altman, jointly awarded a Nobel Prize in 1989, discovered that certain RNAs, now known as ribozymes, showed enzymatic activity. Cech showed that a noncoding sequence (intron) in the small subunit rRNA of protozoans, which had to be removed before the rRNA was functional, can excise itself from a much longer precursor RNA molecule and rejoin the two ends in an autocatalytic reaction. Altman showed that the RNA component of an RNA protein complex called ribonuclease P can cleave a precursor tRNA to generate a mature tRNA. In addition to self-splicing RNAs similar to the one discovered by Cech, artificial RNAs have been made that show a variety of catalytic reactions. It is now widely held that there was a stage during evolution when only RNA catalysed and stored genetic information. This period, sometimes called "the RNA world," is believed to have preceded the function of DNA as genetic material.

Antisense RNAs

Most antisense RNAs are synthetically modified derivatives of RNA or DNA with potential therapeutic value. In nature, antisense RNAs contain sequences that are the complement of the normal coding sequences found in mRNAs (also called sense RNAs). Like mRNAs, antisense RNAs are single-stranded, but they cannot be translated into protein. They can inactivate their complementary mRNA by forming a double-stranded structure that blocks the translation of the base sequence. Artificially introducing antisense RNAs into cells selectively inactivates genes by interfering with normal RNA metabolism.

Viral Genomes

Many viruses use RNA for their genetic material. This is most prevalent among eukaryotic viruses, but a few prokaryotic RNA viruses are also known. Some common examples include poliovirus, human immunodeficiency virus (HIV), and influenza virus, all of which affect humans, and tobacco mosaic virus, which infects plants. In some viruses the entire genetic material is encoded in a single RNA molecule, while in the segmented RNA viruses several RNA molecules may be present. Many RNA viruses such as HIV use a specialized enzyme called reverse transcriptase that permits replication of the virus through a DNA intermediate. In some cases this DNA intermediate becomes integrated into the host chromosome during infection; the virus then exists in a dormant state and effectively evades the host immune system.

Other RNAs

Many other small RNA molecules with specialized functions are present in cells. For example, small nuclear RNAs (snRNAs) are involved in RNA splicing, and other small RNAs that form part of the enzymes telomerase or ribonuclease P are part of ribonucleoprotein particles. The RNA component of telomerase contains a short sequence that serves as a template for the addition of small strings of oligonucleotides at the ends of eukaryotic chromosomes. Other RNA molecules serve as guide RNAs for editing, or they are complementary to small sections of rRNA and either direct the positions at which methyl groups need to be added or mark U residues for conversion to the isomer pseudouridine.

RNA Processing

Cleavage

Following synthesis by transcription, most RNA molecules are processed before reaching their final form. Many rRNA molecules are cleaved from much larger transcripts and may also be methylated or enzymatically modified. In addition, tRNAs are usually formed as longer precursor molecules that are cleaved by ribonuclease P to generate the mature 52 end and often have extra residues added to their 32 end to form the sequence CCA. The hydroxyl group on the ribose ring of the terminal A of the 32 -CCA sequence acts as the amino acid acceptor necessary for the function of RNA in protein building.

Splicing

In prokaryotes the protein coding sequence occupies one continuous linear segment of DNA. However, in eukaryotic genes the coding sequences are frequently "split" in the genome—a discovery reached independently in the 1970s by Richard J. Roberts and Phillip A. Sharp, whose work won them a Nobel Prize in 1993.

The segments of DNA or RNA coding for protein are called exons, and the noncoding regions separating the exons are called introns. Following transcription, these coding sequences must be joined together before the mRNAs can function. The process of removal of the introns and subsequent rejoining of the exons is called RNA splicing.

Each intron is removed in a separate series of reactions by a complicated piece of enzymatic machinery called a spliceosome. This machinery consists of a number of small nuclear ribonucleoprotein particles (snRNPs) that contain small nuclear RNAs (snRNAs).

RNA Editing

Some RNA molecules, particularly those in protozoan mitochondria, undergo extensive editing following their initial synthesis. During this editing process, residues are added or deleted by a posttranscriptional mechanism under the influence of guide RNAs. In some cases as much as 40 percent of the final RNA molecule may be derived by this editing process, rather than being coded directly in the genome. Some examples of editing have also been found in mRNA molecules, but these appear much more limited in scope.

Nucleic Acid Metabolism

DNA Metabolism

Replication, repair, and recombination—the three main processes of DNA metabolism—are carried out by specialized machinery within the cell. DNA must be replicated accurately in order to ensure the integrity of the genetic code. Errors that creep in during replication or because of damage after replication must be repaired. Finally, recombination between genomes is an important mechanism to provide variation within a species and to assist the repair of damaged DNA. The details of each process have been worked out in prokaryotes, where the machinery is more streamlined, simpler, and more amenable to study. Many of the basic principles appear to be similar in eukaryotes.

Replication

Basic Mechanisms

DNA replication is a semiconservative process in which the two strands are separated and new complementary strands are generated independently, resulting in two exact copies of the original DNA molecule. Each copy thus contains one strand that is derived from the parent and one newly synthesized strand. Replication begins at a specific point on a chromosome called an origin, proceeds in both directions along the strand, and ends at a precise point. In the case of circular chromosomes, the end is reached automatically when the two extending chains meet, at which point specific proteins join the strands. DNA polymerases cannot initiate replication at the end of a DNA strand; they can only extend preexisting oligonucleotide fragments called primers. Therefore, in linear chromosomes, special mechanisms initiate and terminate DNA synthesis to avoid loss of information. The initiation of DNA synthesis is usually preceded by synthesis of a short RNA primer by a specialized RNA polymerase called primase. Following DNA replication, the initiating primer RNAs are degraded.

The two DNA strands are replicated in different fashions dictated by the direction of the phosphodiester bond. The leading strand is replicated continuously by adding individual nucleotides to the 32 end of the chain. The lagging strand is synthesized in a discontinuous manner by laying down short RNA primers and then filling the gaps by DNA polymerase, such that the bases are always added in the 52 to 32 direction. The short RNA fragments made during the copying of the lagging strand are degraded when no longer needed. The two newly synthesized DNA segments are joined by an enzyme called DNA ligase. In this way, replication can proceed in both directions, with two leading strands and two lagging strands proceeding outward from the origin.

Enzymes of Replication

DNA polymerase adds single nucleotides to the 32 end of either an RNA or a DNA molecule. In the prokaryote *E. coli*, there are three DNA polymerases; one is responsible for chromosome replication, and the other two are involved in the resynthesis of DNA during damage repair. DNA polymerases of eukaryotes are even more complicated. In human cells, for instance, more than five different DNA polymerases have been characterized. Separate polymerases catalyse the synthesis of the leading and lagging strands in human cells, and a separate polymerase is responsible for replication of mitochondrial DNA. The other polymerases are involved in the repair of DNA damage.

A number of other proteins are also essential for replication. Proteins called DNA helicases help to separate the two strands of DNA, and single-stranded DNA binding proteins stabilize them during opening prior to being copied. The opening of the DNA helix introduces considerable strain in the form of supercoiling, a movement that is subsequently relaxed by enzymes called topoisomerases. A special RNA polymerase called primase synthesizes the primers needed at the origin to begin transcription, and DNA ligase seals the nicks formed between individual fragments.

The ends of linear eukaryotic chromosomes are marked by special sequences called telomeres that are synthesized by a special DNA polymerase called telomerase. This enzyme contains an RNA component that serves as a template for the exact sequence found at the ends of chromosomes. Multiple copies of a short sequence within the telomerase-associated RNA are made and added to the telomere ends. This has the effect of preventing shortening of the DNA chain that would otherwise occur during replication.

Single-stranded viral genomes, mitochondrial genomes, and some viral genomes are replicated in specialized ways. Several viruses such as adenovirus use a nucleotide covalently bound to a protein as a primer, and the protein remains covalently bound to the DNA after replication. Many single-stranded viruses use a rolling circle mechanism of replication whereby a double-stranded copy of the virus is first made. The replicating machinery then copies the nonviral strand in a continuous fashion, generating long single-stranded DNA from which full-length viral DNA strands are excised by specialized nucleases.

Recombination

Recombination is the principal mechanism through which variation is introduced into populations. For example, during meiosis, the process that produces sex cells (sperm or eggs), homologous chromosomes— one derived from the mother and the equivalent from the father— become paired, and recombination, or crossing-over, takes place. The two DNA molecules are fragmented, and similar segments of the chromosome are shuffled to produce two new chromosomes, each being a mosaic of the originals. The pair separates so that each sperm or egg receives just one of the shuffled chromosomes. When sperm and egg fuse, the normal set of two copies of each chromosome is restored.

There are two forms of recombination, general and site-specific. General recombination typically involves cleavage and rejoining at identical or very similar sequences. In site-specific recombination, cleavage takes place at a specific site into which DNA is usually inserted. General recombination occurs among viruses during infection, in bacteria during conjugation, during transformation whereby DNA is directly introduced into cells, and during some types of repair processes. Site-specific recombination is frequently involved in the parasitic distribution of DNA segments throughout genomes. Many viruses, as well as special segments of DNA called transposons, rely on site-specific recombination to multiply and spread.

General Recombination

General recombination, also called homologous recombination, involves two DNA molecules that have long stretches of similar base sequences. The DNA molecules are nicked to produce single strands; these subsequently invade the other duplex, where base pairing leads to a four-stranded DNA structure. The cruciform junction within this structure is called a Holliday junction, named after Robin Holliday, who proposed the original model for homologous recombination in 1964. The Holliday junction travels along the DNA duplex by "unzipping" one strand and reforming the hydrogen bonds on the second strand. Following this branch migration, the two duplexes can be nicked again, allowing them to separate. Finally, the nicks are repaired by DNA ligase. The result is two DNA duplexes in which the segment between the two nicks has been replaced. The enzymes involved in recombination have been characterized best in the prokaryote *E. coli*. A key enzyme is RecA, which catalyzes the strand invasion process. RecA coats single-stranded DNA and facilitates its pairing with a double-stranded DNA molecule containing the same sequence, which produces a loop structure.

Another protein, known as RecBC, is important for the recombination process. Functioning at free ends of DNA, RecBC catalyzes an unwinding-rewinding reaction as it traverses the length of the molecule. Since unwinding is faster than rewinding, a loop is produced behind the enzyme that facilitates subsequent pairing with another DNA molecule. A number of other proteins are also important for recombination, including single-stranded DNA binding proteins that stabilize single-stranded DNA, DNA polymerase to repair any gaps that might be formed, and DNA ligase to reseal the nicks after recombination is complete. The details of eukaryotic recombination are expected to parallel those found in *E. coli*, although the highly compact chromatin structure in eukaryotes makes the process more complicated.

It is important to note that the initial product of recombination between two regions of DNA that are similar but not identical will be a "heteroduplex"—that is, a molecule in which mismatched bases will be present at some positions in the helix. Thus, in the specialized recombination that takes place during meiosis, one round of replication is necessary before the mosaic chromosomes produced by recombination are properly matched. Enzymes are present in cells that specifically recognize and repair mismatches, so that the initial products of recombination can sometimes be repaired before they are replicated. In such cases the final products of replication will not be true reciprocal events, but rather one of the original parental molecules will appear to have been maintained to the exclusion of the other—a process called gene conversion.

Recombination also functions occasionally to repair lesions in DNA. If one chromosome of a pair becomes irreversibly damaged, the information from the other chromosome can be copied and inserted by recombination to provide a correct replacement of the damaged section. The key idea here is that sequences flanking the damage from a sister chromosome can base-pair with the corresponding sequences on the damaged chromosome, thus allowing replication to copy the correct sequence and repair the lesion.

Site-specific Recombination

Site-specific recombination involves very short specific sequences that are recognized by proteins. Long DNA sequences such as viral genomes, drug-resistance elements, and regulatory sequences such as the mating type locus in yeast can be inserted, removed, or inverted, having profound regulatory effects. More than any other mechanism, site-specific recombination is responsible for reshaping genomes. For example, the genomes of many higher organisms, including plants and humans, show evidence that transposable elements have been constantly inserted throughout the genome and even into one another from time to time.

One example of site-specific recombination is the integration of DNA from bacteriophage λ into the chromosome of *E. coli*. In this reaction, bacteriophage λ DNA, which is a linear molecule in the normal phage, first forms a circle and then is cleaved by the enzyme λ-integrase at a specific site called the phage attachment site. A similar site on the bacterial chromosome is cut by integrase to give ends with the identical extension. Because of the complementarity between these two ends, they can be rejoined so that the original circular λ chromosome is inserted into the chromosome of the *E. coli* bacterium. Once integrated, the phage can be held in an inactive state until signals are generated that reverse the process, allowing the phage genome to escape and resume its normal life cycle of growth and spread into other bacteria. This site-specific recombination process requires only λ-integrase and one host DNA binding protein called the integration host factor. A third protein, called excisionase, recognizes the hybrid sites formed on integration and, in conjunction with integrase, catalyzes an excision process whereby the λ chromosome is removed from the bacterial chromosome.

A similar but more widespread version of DNA integration and excision is exhibited by the transposons, the so-called jumping genes. These elements range in size from fewer than 1,000 to as many as 40,000 base pairs. Transposons are able to move from one location in a genome to another, as first discovered in corn (maize) during the 1940s and '50s by Barbara McClintock, whose work won her a Nobel Prize in 1983. Most, if not all, transposons encode an enzyme called transposase that acts much like λ-integrase by cleaving the ends of the transposon as well as its target site.

Transposons differ from bacteriophage λ in that they do not have a separate existence outside of the chromosome but rather are always maintained in an integrated site. Two types of transposition can occur—one in which the element simply moves from one site in the chromosome to another and a second in which the transposon is replicated prior to moving. This second type of transposition leaves behind the original copy of the transposon and generates a second copy that is inserted elsewhere in the genome. Known as replicative transposition, this process is the mechanism responsible for the vast spread of transposable elements in many higher organisms.

The simplest kinds of transposons merely contain a copy of the transposase with no additional genes. They behave as parasitic elements and usually have no known associated function that is advantageous to the host. More often, transposable elements have additional genes associated with them—for example, antibiotic resistance factors. Antibiotic resistance typically occurs when an infecting bacterium acquires a plasmid that carries a gene encoding resistance to one or more antibiotics.

Typically, these resistance genes are carried on transposable elements that have moved into plasmids and are easily transferred from one organism to another. Once a bacterium picks up such a gene, it enjoys a great selective advantage because it can grow in the presence of the antibiotic. Indiscriminate use of antibiotics actually promotes the buildup of these drug-resistant plasmids and strains.

Repair

It is extremely important that the integrity of DNA be maintained in order to ensure the accurate workings of a cell over its lifetime and to make certain that genetic information is accurately passed from one generation to the next. This maintenance is achieved by repair processes that constantly monitor the DNA for lesions and activate appropriate repair enzymes. As described in the section General recombination, serious lesions in DNA such as pyrimidine dimers or gaps can be repaired by recombination mechanisms, but there are many other repair mechanisms.

One important mechanism is that of mismatch repair, which has been studied extensively in *E. coli*. The system is directed by the presence of a methyl group within the sequence GATC on the template strand. Comparable systems for mismatch repair also operate in eukaryotes, though the template strand is not marked by methyl groups. In fact, lesions within the genes for human mismatch repair systems are known to be responsible for many cancers. Loss of the mismatch repair system allows mutations to build up quickly and eventually to affect the genes that cause cells to divide. As a result, cells divide in an uncontrolled manner and become cancerous.

Once replication is complete, the most common kind of damage to nucleic acids is one in which the normal A, C, G, and T bases are changed into chemically modified bases that usually differ significantly from their natural counterparts. The only exceptions are the deamination of cytosine to uracil and the deamination of 5-methylcytosine to thymine.

In these cases the product is a G:U or G:T mismatch. Specific enzymes called DNA glycosylases can recognize uracil in DNA or the thymine in a G:T mismatch and can selectively remove the base by cleaving the bond between the base and the deoxyribose sugar. Many of these enzymes are specific for the different chemically modified bases that may be present in DNA.

Another common means of repairing DNA lesions is by an excision repair pathway. Enzymes recognize damage within DNA, probably by detecting an altered conformation of DNA, and then nick the strand on either side of the lesion, allowing a small single-stranded DNA to be excised. DNA polymerase and DNA ligase then repair the single-stranded gap. In all of these systems, the presence of an abnormal base signifies which strand is to be repaired, and the complementary strand is used as the template to ensure the accuracy of repair.

RNA Metabolism

RNA provides the link between the genetic information encoded in DNA and the actual workings of the cell. Some RNA molecules such as the rRNAs and the snRNAs (described in the section Types of RNA) become part of complicated ribonucleoprotein structures with specialized roles in the cell. Others such as tRNAs play key roles in protein synthesis, while mRNAs direct the synthesis of proteins by the ribosome. Three distinct phases of RNA metabolism occur.

First, selected segments of the genome are copied by transcription to produce the precursor RNAs.

Second, these precursors are processed to become functionally mature RNAs ready for use. When these RNAs are mRNAs, they are then used for translation.

Third, after use the RNAs are degraded, and the bases are recycled. Thus, transcription is the process where a specific segment of DNA, a gene, is copied into a specific RNA that encodes a single protein or plays a structural or catalytic role. Translation is the decoding of the information within mRNA molecules that takes place on a specialized structure called a ribosome.

There are important differences in both transcription and translation between prokaryotic and eukaryotic organisms.

Transcription

Small segments of DNA are transcribed into RNA by the enzyme RNA polymerase, which achieves this copying in a strictly controlled process. The first step is to recognize a specific sequence on DNA called a promoter that signifies the start of the gene. The two strands of DNA become separated at this point, and RNA polymerase begins copying from a specific point on one strand of the DNA using a ribonucleoside 52 -triphosphate to begin the growing chain.

Additional ribonucleoside triphosphates are used as the substrate, and, by cleavage of their high-energy phosphate bond, ribonucleoside monophosphates are incorporated into the growing RNA chain. Each successive ribonucleotide is directed by the complementary base pairing rules of DNA.

Thus, a C in DNA directs the incorporation of a G into RNA, G is copied into C, T into A, and A into U. Synthesis continues until a termination signal is reached, at which point the RNA polymerase drops off the DNA, and the RNA molecule is released. In some cases this RNA molecule is the final mRNA.

In other cases it is a pre-mRNA and requires further processing before it is ready for translation by the ribosome. Ahead of many genes in prokaryotes, there are signals called "operators" where specialized proteins called repressors bind to the DNA just upstream of the start point of transcription and prevent access to the DNA by RNA polymerase.

These repressor proteins thus prevent transcription of the gene by physically blocking the action of the RNA polymerase. Typically, repressors are released from their blocking action when they receive signals from other molecules in the cell indicating that the gene needs to be expressed. Ahead of some prokaryotic genes are signals to which activator proteins bind that positively induce transcription.

Transcription in higher organisms is more complicated. First, the RNA polymerase of eukaryotes is a more complicated enzyme than the relatively simple five-subunit enzyme of prokaryotes. In addition, there are many more accessory factors that help to control the efficiency of the individual promoters.

These accessory proteins are called transcription factors and typically respond to signals from within the cell that indicate whether transcription is required. In many human genes, several transcription factors may be needed before transcription can proceed efficiently. A transcription factor can cause either repression or activation of gene expression in eukaryotes.

During transcription, only one strand of the DNA is usually copied. This is called the template strand, and the RNA molecules produced are single-stranded. The DNA strand that would correspond to the mRNA is called the coding or sense strand, and it is not unusual for this to change from one gene to the next. In eukaryotes the initial product of transcription is called a pre-mRNA, which is extensively spliced before the mature mRNA is produced, ready for translation by the ribosome.

Translation

The process of translation uses the information present in the nucleotide sequence of mRNA to direct the synthesis of a specific protein for use by the cell. Translation takes place on the ribosomes—complex particles in the cell that contain RNA and protein.

In prokaryotes the ribosomes are loaded onto the mRNA while transcription is still ongoing. Near the 52 end of the mRNA, a short sequence of nucleotides signals the starting point for translation. It contains a few nucleotides called a ribosome binding site, or Shine-Dalgarno sequence. In *E. coli* the tetranucleotide GAGG is sufficient to serve as a binding site. This typically lies five to eight bases upstream of an initiation codon.

The mRNA sequence is read three bases at a time from its 52 end toward its 32 end, and one amino acid is added to the growing chain from its respective aminoacyl tRNA, until the complete protein chain is assembled.

Translation stops when the ribosome encounters a termination codon, normally UAG, UAA, or UGA. Special release factors associate with the ribosome in response to these codons, and the newly synthesized protein, tRNAs, and mRNA all dissociate. The ribosome then becomes available to interact with another mRNA molecule.

In eukaryotes the essence of protein synthesis is the same, although the ribosomes are more complicated. As with prokaryotic initiation, the signal sequence interacts with the 32 end of the small subunit rRNA during formation of the initiation complex.

The issue of fidelity is important during protein synthesis, but it is not as crucial as fidelity during replication. One mRNA molecule can be translated repeatedly to give many copies of the protein.

When an occasional protein is mistranslated, it usually does not fold properly and is then degraded by the cellular machinery. However, proofreading mechanisms exist within the ribosome to ensure accurate pairing between the codon in the mRNA and the anticodon in the tRNA.

One of the crowning achievements of molecular biology was the elucidation during the 1960s of the genetic code. Principals in this effort were Har G. Khorana and Marshall W. Nirenberg, who shared a Nobel Prize in 1968. Khorana and Nirenberg used artificial templates and protein synthesizing systems in the test tube to determine the coding potential of all 64 possible triplet codons. The key feature of the genetic code is that the 20 amino acids are encoded by 61 codons. Thus, there is degeneracy in the code such that one amino acid is often specified by more than one codon.

In the case of serine and leucine, six codons can be used for each. Among organisms that have been examined in detail, the code appears to be almost universal, from bacteria through archaea to eukaryotes. The known exceptions are found in the mitochondria of humans and many other organisms as well as in some species of bacteria. The structure within the genetic code whereby many amino acids are uniquely coded by the first two bases of the codon strongly suggests that the code has itself evolved from a more primitive code involving 16 dinucleotides. How the individual amino acids became associated with the different codons remains a matter of speculation.

8

Manipulating DNA

The Manipulating DNA section covers all the things that people are able to do with/to DNA, as well as the things still in the realm of science fiction. The following is links and information about each of the sections listed in the navigation frame on the left.

Cloning

This area explores experimentation done involving Cloning, including Dolly and Missy. The idea of cloning has been considered since before the discovery of DNA. Cloning may give us insights into what makes us individuals, and how much of our personality and behaviour is based on our genetics. It may also destroy our sense of individuality and the value of life.

It seems that every week, newspapers report on new advances in the science of cloning. Everybody knows about Dolly the cloned sheep, but few people know all the details about cloning, including the fact that scientists have been working on it for over 100 years.

Cloning in Nature

Cloning has been going on in the natural world for thousands of years. A clone is simply one living thing made from another, leading to two organisms with the same set of genes. In that sense, identical twins are clones, because they have identical DNA. Sometimes, plants are self-pollinated, producing seeds and eventually more plants with the same genetic code. Some forests are made entirely of trees originating from one single plant; the original tree spread its roots, which later sprouted new trees. When earthworms are cut in half, they regenerate the missing parts of their bodies, leading to two worms with the same set of genes. However, the ability to intentionally create a clone in the animal kingdom by working on the cellular level is a very recent development.

Early Progress

The first cloned animals were created by Hans Dreisch in the late 1800's. Dreich's original goal was not to create identical animals, but to prove that genetic material is not lost during cell division. Dreich's experiments involved sea urchins, which he picked because they have large embryo cells, and grow independently of their mothers. Dreich took a 2 celled embryo of a sea urchin and shook it in a beaker full of sea water until the two cells separated. Each grew independently, and formed a separate, whole sea urchin.

In 1902, another scientist, embryologist Hans Spemman, used a hair from his infant son as a knife to separate a 2-celled embryo of a salamander, which also grow externally. He later separated a single cell from a 16-celled embryo. In these experiments, both the large and the small embryos developed into identical adult salamanders.

Spemman went on to propose what he called a "fantastical experiment" — to remove the genetic material from an adult cell, and use it to grow another adult. In this way, he theorized, he would be able to prove that no genetic material was lost as cells grew and divided.

New Advances

There were no major advances in cloning until November of 1951, when a team of scientists in Philadelphia working at the lab of Robert Briggs cloned a frog embryo. This team did not simply break off a cell from an embryo, however. They took the nucleus out of a frog embryo cell and used it to replace the nucleus of an unfertilized frog egg cell, completing the "fantastical experiment" of nearly 50 years before. Once the egg cell detected that it had a full set of chromosomes, it began to divide and grow. This was the first time that this process, called nuclear transplant, was ever used, and it continues to be used today, although the method has changed slightly.

False Hopes

In 1977, a German scientist shocked the world, claiming to have cloned three mice from embryos. Although embryos had been cloned before, no one had been able to do the experiment with mice because the cells were so small and the tools so large that the cells were traumatized and would eventually die after a few divisions. He instantly became famous, telling the world how he cloned his mice. However, he refused to actually demonstrate any of his techniques, and when other scientists couldn't replicate his work, he came under suspicion. He was challenged — repeat his work or be discredited. He accepted.

He claimed to work nights and mornings when no one was around, but the equipment was never disturbed. He showed off his mouse embryos' growth daily, even though a malfunction in the water purification system left other scientists at his lab unable to grow other embryos.

Later, in his cabinet, test tubes were found with mouse embryos in them, each at a different stage of development. Most scientists do not believe that this scientist was ever able to clone adult mice. In 1978, a science fiction writer published a book claiming that a millionaire (known to the readers only as Max) had come to him because of his connections as a writer, and asked the him to arrange for Max to be cloned.

The author eventually agreed, as the story goes, and Max was cloned. The book was ranked in the Top 10 list of popular books. Scientists who read his book, however, noticed discrepancies between the book and scientific data. One man who was quoted in the book was angry enough to sue. The publisher admitted that the book was a hoax, but the author maintains his claim to this day.

Within these two years, two front-page advances in cloning were discovered to be, most likely, frauds. As a direct result, many scientists began to claim that cloning of mammals was impossible. Funding and interest dropped, and cloning returned to the realm of science fiction for several years.

First Cloned Mammals

A breakthrough came in 1986. Two teams, working independently but using nearly the same method, each on opposites side of the Atlantic, announced that they had cloned a mammal. One team was led by Steen Willadsen in England, which cloned a sheep's embryo. The other team was led by Neal First in America, which cloned a cow's embryo.

Many advances were made during the course of these experiments, including progress in keeping tissue alive in lab conditions. However, neither team believed that it was possible to clone from an adult's differentiated cells. With no progress in sight, the prospect of cloning fell by the wayside, and little research was done on the matter.

Dolly

Ian Wilmut at the Roslin Institute in Scotland was assigned to a project in 1986. His goal was to create a sheep that produced a certain chemical in its milk. He chose to alter adult cells, which held up well in laboratory conditions, and then clone them, producing animals with the altered gene all throughout their bodies. He began the paperwork in 1987, and began research in 1990.

One of Wilmut's colleagues, who had experience with cloning from early embryo cells, suggested that the reason so many cloning attempts failed was that the cells were in incompatible stages of life. In one stage, the cells are adding to the DNA, in another, they are proofreading it, and in another, splitting it.

The cells, he theorized, could not always start over. Wilmut's team learned that by starving the cells, they could be forced into what is called the G0 phase, similar to cellular hibernation. This advance increased the survival rate of the cloned cells; Megan and Morag, two lambs, were cloned from sheep embryos. Wilmut's team now realized that differentiation did not matter in cloning. More work was done, and on July 5, 1996, a lamb was born, cloned from a frozen mammary cell from another adult sheep. Wilmut, who names his animals very creatively, named her Dolly after Dolly Parton.

Although Dolly was just a step in a long experiment, the press descended upon the first animal cloned from an adult. The Roslin Institute was overrun with journalists and reporters. However, other scientists were critical — Dolly took 277 tries to create, and other labs were unable to reproduce the results. In addition, it took over a year for the institute to test Dolly's DNA to make sure that it was indeed the same as that of the frozen mammary cells. Science, although temporarily impressed, demanded a better way.

Herd of Mice

Oct 3, 1997, the Honolulu Technique created Cumulina the cloned mouse. She was cloned from cumulus cells (cells which surround developing egg cells) using traditional nuclear transfer. The nucleus was taken from the cumulus cell and implanted in an egg cell from another mouse. The new cell was then treated with a chemical to make it grow and divide. The scientists repeated the process for three generations, yielding over fifty mice that are virtually identical by the end of July, 1998. The Honolulu Technique's success rate of 50:1 is almost six times better than that of the Roslin Institute's success rate, 277:1. As cloning technology improves, more and more applications will be seen in everyday life.

Mainstream Cloning

How much do you love your dog? Is your dog so perfect that you would pay over $2.3 million dollars to have another just like it?

One couple thinks their 11-year-old dog is just such an animal. Wishing to remain anonymous to avoid run-ins with the press, this couple has contracted Texas A&M University to clone their dog, Missy. Scientists are hailing this for its scientific achievement; no dogs have been cloned before because their reproductive system is rather complicated. If the cloning of dogs can be achieved, perhaps exceptional animals like rescue animals can be reproduced.

In addition to the pure scientific appeal of cloning a dog, the attempt to clone Missy has another interesting addition to make to the history of cloning. A private couple wants their dog cloned. They are, of course, spending millions to have her cloned, but consider the possibilities. Could cloning the family pet one day become a normal alternative to buying a new one?

Applications

Reliable cloning can be used to make farming more productive by replicating the best animals. It can make medical testing more accurate by providing test subjects that all react the same way to the same drug. It can allow mass production of genetically altered animals, plants, and bacteria. It may settle once and for all what part of personality is dependent on genetics and what part on environment. In short, it can be beneficial to almost every area of biological science.

How Cloning Works

Cloning in Nature

In nature, clones exist among plants or animals that are able to reproduce asexually. Special traits allow earthworms to regenerate lost body parts and certain plants (like the spider/airplane plant) to propogate through runners. These worms and plants are technically clones, because as broken-off parts of the original, they share the same genetic material as the original. The oldest methods of cloning involve a similar principle — a cell is broken off from a very early embryo, and begins to divide as if it were a fertilized egg. However, it is very difficult to clone adult cells because they are not "in gear" to divide and grow into a new organism. Techniques needed to be developed that could allow the DNA from an adult cell to be removed and placed in a cell that could grow as an embryo.

Nuclear Transfer

Scientists' understanding of what makes an embryo cell different from a liver cell is not complete because the objects of study are so small. However, they do know that a coating of proteins on the cell wall inhibits the DNA that is not neccesary for that cell's particular function. When this coating is present, the cell is said to be differentiated. Although differentiated cells contain all the necessary DNA to produce a new organism, the fact that they have differentiated keeps them from creating an entire new organism. Egg cells, on the other hand, are unique in that they are the only cells capable of forming a new and complete organism.

The secret to cloning lies in removing an egg cell's nucleus and replacing it with the nucleus of a donor cell. The egg cell, which has a "clean slate", and is primed to use the information in the DNA to create a new organism, now has the DNA of the donor organism. However, this transfer is a very difficult process, more difficult in some cases than others.

Obstacles to Cloning

The first animals cloned using the modern method of nuclear transfer were frog embryos, cloned in 1951. The reason frog embryos were cloned long before anything mammalian is twofold. First, the cells of an 8-celled frog embryo are huge compared to the cells of a mammal embryo; they can, in fact, be seen with the naked eye, and are therefore much easier to work with.

Secondly, frog embryo cells grow in lab conditions much better than mammal embryo cells. This is because frog embryos are supposed to grow in water, without any added nutrition or outside assistance. Mammals, however, take a long time to mature, and require the shelter and nourishment of a mother both before and after birth.

Because the clone has to grow from a single cell to an entire organism in the lab, it was easier to do initial experiments on amphibians. Frogs were chosen because, as frequent subjects in lab tests, they are very well known to scientists.

Although happy about the cloned frogs, scientists wanted to see a cloned mammal; humans are mammals, so almost anything learned about other mammals can be applied to humans.

Unfortunately, because of the difficulties associated with mammal cloning, this was a difficult task. Techniques to make this task possible were developed at the labs of Neal First and Steen Willadsen, who worked independently to produce the first mammals cloned from embryos. Once the nucleus is transferred from the donor cell to the egg cell, the egg cell is put in a nutrient solution until it divides into an eight-celled embryo.

When the cell passes this point, the individual cells begin to differentiate; one cell may eventually form the head, and another the torso. As this point is reached, the scientists move quickly to put the embryo into the removed and artificially sustained oviduct of a sheep or other animal.

The oviduct functions as an incubator to nurture the embryo thruough this critical period. Once the embryo is large enough to be able to withstand the procedure, it is placed in a living surrogate mother where it will be carried to term, nourished by an umbilical cord.

With all these steps, it is no wonder that the success rate for cloning is so low; embryos die at each step along the way until only a few are carried to term and born.

Variations on a Theme

Of course, there are different approaches to all steps in the process. Ian Wilmut, instead of transferring the nucleus from one cell to another, fused an enucleated (nucleusless) egg cell with an adult cell to form the first cell in the embryo.

The Honolulu method centres around the cells chosen to be DNA donors. However, the principle behind cloning is always the same - give an egg cell, which is ready to divide and grow into a new organism, the nucleus of an old cell. With the new set of instructions, the egg cell begins its journey to become an adult clone.

Future of Cloning

Consider Some of the Possible Applications of Cloning

Cloning is truly a monumental accomplishment. But what can we do with it? Very few see a future society of cloned humans. Mad scientists with armies of cloned zombies are highly improbable due to the current cost, failure rate, and time involved.

However, there are some practical applications currently being considered. Which are possible and which are doomed to remain in the realm of science fiction? Only the future will tell.

Most Prevalent Application

What many scientists forsee as the most probable and widespread application of cloning is the mass production of genetically engineered animals. Scientists have to be very careful about protecting the altered portion of the genome; if the new/altered gene is damaged, the animal carrying the damaged gene will not be as useful to the scientists.

Because natural breeding can lead to the decay of the new/altered gene, cloning is the best option, because the original DNA is used over and over to make new animals. In fact, mammalian cloning was only pursued as a step toward copying genetically altered cells.

Farming

The average dairy cow puts out roughly 15,000 gallons of milk a year. However, there are certain, special cows that can make up to 45,000 gallons a year. Selectively breeding for this trait is nearly impossible given the complexity of the genetics governing milk production. However, if scientists clone these exceptional cows, the profit gained from increased milk outweighs the money spent on a cloned cow.

Infertility Treatment

While it is unlikely that the cloning of humans will ever become standard practice, it is also unlikely that it will never be attempted. Many people see cloning as a way to provide children for those couples who cannot have them naturally. While there is always someone willing and able to pay the price, is the cloning of humans morally right?

Jurassic Park

Scientists have told us that it would be virtually impossible to find an intact cell from a extinct dinosaur. However, at New Zealand's

University of Otago, scientists are trying to clone the extinct moa, once the world's largest bird. The plan was to take DNA from its leg bone and implant it into a chicken egg to grow.

The scientists were then going to breed the new moa with an ostrich or an emu to create a new giant bird. Research was stopped by the Ngai Tahu Maori tribe, which claims to own the DNA because they were sole owners of the land when the bird became extinct around 1500.

In 1992, a herd of cows that was damaging the ecological balance of New Zealand's Enderby Island was almost completely destroyed. All that remained was the frozen sperm of ten bulls, and one living female, Lady. Because Lady is so old, all attempts to make her pregnant using the frozen sperm have failed. Therefore, Lady was cloned. On July 31, 1998, Elsie (L.C., Lady's Clone) was delivered by Casarean section, is the hope of preserving the nearly-extinct Enderby Island cows.

Organs for Transplant

Another option open to the future is the cloning of specific organs for transplant. Each year, many people die, unable to find a suitable organ donor. In addition, those who do find donors many times have to take anti-rejection drugs for the rest of their lives. If scientists can find a way to force cells to differentiate to become a failing organ, they should be able to grow those cells into a working, adult organ. Because the new organ would be an exact match for the patient, there would be no need for anti-rejection drugs. Many predict that the first organ to be cloned in this way may be bone marrow, because it is a liquid organ and has no shape. However, it is conceivable that solid organs would be able to be cloned outside a body as well.

Ethics of Cloning

We can... but should we?

Cloning, simply defined, is creating a new organism that shares the same genetic code as another. Each step taken toward making cloning quicker, better, and cheaper brings these ethical questions further from the realm of the hypothetical and closer to the realm of fact.

Cloning Endangered/Extinct Species

All endangered species have one thing in common — there's not many animals left. A lack of animals leads to inbreeding, and a lack of genetic diversity. Researchers are always careful when trying to reestablish endangered species to maintain genetic diversity by creating different families and trying to inbreed as little as possible. If the scientists just let the animals breed without control, there will be almost no genetic diversity, which will lead to recessive traits being expressed. Usually, this results in diseases and deformities.

While cloning offers a short-term answer to the problem of "not enough animals", in the long term, all the offspring from the clones are genetically brothers and sisters. This will likely destroy genetic diversity and lead to inbred, weak, disease ridden animals who could not survive in the wild.

Even if the animals were healthy enough to be released into the wild, in their weakened state, they would be easy prey, leading to more predators. In the end, they would become extinct and would have upset nature's balance by introducing more predators. The morality of cloning endangered species hinges on how many animals are left, and what their collective genetic condition is.

The same trouble would be compounded if science tried to bring back an endangered species like the dodo bird. It would be so expensive to find enough intact dodo cells from different birds, that to make a flock of unique birds would be an impracticality. The best science can offer is the same cloned dodo at every paying zoo. Whether or not you think this ethical hinges on the question of whether it is ethical to create an animal for the sole purpose of captivity.

Cloning Humans

It would clearly be unethical to clone humans and distribute them about for the purpose of a gigantic twin study, to determine once and for all what is nature and what is nurture. It would be useless for an millionare to clone himself in hopes of everlasting life; the clone would posess none of his memories or experiences. Bill Gates' clone might turn into an average person, a delinquent, a ditch digger, or the secretary of the Department of Justice, depending on the clone's life experience.

There is really only one apparent and plausible reason for the widespread cloning of humans, and that is for parents who can't have children naturally. How you feel about cloning humans really depends on how you feel about the value of human life, and when it starts. Dolly took 277 tries; assuming that cloning a human would take 277 tries, what of the 276 human embryos that died? If they are just embryos, then it's probably okay. If they were humans with minds, souls, and destinies, then it was wrong to try to clone a child in the first place. Cloning, it seems, offers the same ethical dilemmas as does abortion and current methods of in vitro fertilization.

But there is a second problem on top of the consideration of the other embryos — the life of the clone. Assuming that the technique is commonplace (the lives of the first fifty or so will probably be filled with doctor visits, disgruntled family members, and possibly taunting classmates), a cloned child would have a set of problems in addition to the ones that normal children face.

What would be the relationship between the clone and the "DNA donor"? Sibling? Child? Most children have problems with their parents; a clone might reflect those feelings back on himself, because he is identical to them. There is a whole list of questions that prospective parents would have to consider before trying such an endeavor; is it really worth it?

Core Issue

The core issue behind cloning seems to have to do with the relationship between genetic uniqueness and personal individuality. Animals may or may not have their own personalities and sense of individuality, and if they exist, they may differ from species to species, so there may never be a resolution to the question of cloning animals.

But humans are a different matter entirely. Humans are unique not because of their body build or genetic makeup or life experience, but because of the unique contribution they each have to make to the world. The morality of human cloning lies not in the cloning process itself, but in societal reaction to him.

If society can keep itself from branding the clone as a duplicate person and limiting him to his predecessor's abilities; if instead the clone is accepted just as a normal human being who is unique because of his unique contribution he has to make to the world, then cloning of humans may be an option.

Genetic Engineering

Learn what genetic engineering can accomplish and how it is done. If genetic engineering is defined as changing an organism's DNA to make it more beneficial, genetic engineering has been going on for a very, very long time in the form of selective breeding. However, actually going into a cell and changing its genome by inserting or removing DNA is a very new technology.

Ancient History

Selective breeding has been going on for countless generations. In fact, it is even mentioned in the Bible (Genesis 30:25 - 43). In the account, Jacob was employed as a shepherd under his father-in-law Laban. Instead of receiving wages, Jacob received the black, streaked, and spotted sheep, and Laban kept all the white sheep.

Jacob craftily arranged for his black sheep to mate with Laban's white sheep, producing streaked and spotted sheep. Jacob did so well with this scheme that Laban's family began to get mad at Jacob, and he eventually had to leave.

Difficulties

Selective breeding is effective enough if the goal is to maintain or gradually improve a group of animals. Over the decades, selective breeding has brought us improved strains of cattle and specialized breeds of dogs. However, these advances have taken hundreds of years to effect. In addition to the time concerns, it is often impossible to know which traits will be transferred to the offspring.

Limits

Selective breeding is a long, tedious process that has its limits. It is impossible through selective breeding to mix traits from two totally different species. If a junkyard owner wanted a guard dog that could squirt ink like an octopus, he would be unable to create such an animal. It is physically impossible, because the genetics of life are such that traits from two different organisms cannot be mixed. That is where genetic engineering comes in.

The Progress

Modern genetic engineering began in 1973 when Herbert Boyer and Stanley Cohen used enzymes to cut a bacteria plasmid and insert another strand of DNA in the gap. Both bits of DNA were from the same type of bacteria, but this milestone, the invention of recombinant DNA technology, offered a window into the previously impossible — the mixing of traits between totally dissimilar organisms. To prove that this was possible, Cohen and Boyer used the same process to put a bit of frog DNA into a bacteria.

Since 1973, this technology has been made more controllable by the discovery of new enzymes to cut the DNA differently and by mapping the genetic code of different organisms.

Now that we have a better idea of what part of the genetic code does what, we have been able to make bacteria that produce human insulin for diabetics (previously came from livestock), as well as EPO for people on kidney dialysis (previously came from urine of people in third world countries with ringworm).

In 1990, a young child with an extremely poor immune system recieved genetic therapy. Some of her white blood cells were genetically manipulated and re-introduced into her bloodstream while she watched Sesame Street. These new cells have taken over for the original, weak white cells, and her immune system now works properly. Although relatively few people have had their cells genetically altered, these advances have made the prospect of mainstream genetic medicine seem more likely.

The Promise

Genetic engineers hope that with enough knowledge and experimentation, it will be possible in the future to create "made-to-order" organisms. This will lead to new innovations, possibly including custom bacteria to clean up chemical spills, or fruit trees that bear different kinds of fruit in different seasons. Any trait occurring in nature can theoretically be mixed with any other to form a totally new organism that would not otherwise occur in nature.

Current Status

As of late summer of 1998, scientists are able to add simple traits to organisms. They cannot create custom-made animals. They cannot always predict how traits will interact. Before phenomenally new advances can be made, scientists have to learn how to affect cells' DNA with pin-point accuracy, without affecting other traits. Advances like genetic correction for nearsightedness are a long way off. The power of science is limited to knowledge about genetics, gene locations, and trait interactions, but as knowledge grows, so will scientists' abilities to manipulate life.

How Genetic Engineering Works

On most TV shows, genetic engineering is a very simple process. Just point the device at the person, press the button, and automatically, the patient's DNA is all changed, and the scientist, who often has little idea how to use the machine, is thrilled to death that it worked right the first time. While this makes good fiction, it is hardly reality. Genetic engineering, while the results are often unexpected, is a very meticulous, very exact process with a clear goal in mind.

Cutting the DNA

Genetic engineering has three basic parts. First, the DNA is cut in the desired place using restriction enzymes. Each different type of restriction enzyme "seeks out" and cuts DNA at a spot marked by a different sequence of base pairs. One restriction enzyme may cut the DNA at every "AATC", for example, while another cuts all "ATG" sequences.

The DNA is cut in such a way that one helix is a bit longer than the other because of a few extra base pairs. These extra base pairs are called the "sticky end", because they will bond easily to another strand of DNA with the corresponding set of genes. The new genes that will be injected have sticky ends that complement the sticky ends the restriction enzyme leaves.

Inserting New DNA

Once the restriction enzyme has cut the DNA, new DNA is inserted into the cell. This is difficult, because cells by nature do not allow DNA through the cell wall. There are many ways around this difficulty, though. Electroporation involves jolting the cells with a burst of electricity, opening the cell wall pores and allowing DNA to fall into the cell.

Microinjection uses a small glass needle to inject the DNA through the cell wall. A gene gun can be used, which blasts tiny metal fragments coated with DNA through the cell wall and into the cell. DNA can also be shrouded in lipids, fatty molecules which the cell will take in; when the lipid is digested, the DNA is released. These methods work fairly well for prokaryotic cells, which have one chromosome and no nucleus. However, in order for DNA insertion to work consistently and accurately in multicellular organisms with eukaryotic cells, something better is called for.

To find that something better, science turned to the natural world. They realized that a retrovirus, a form of a virus, was just what they needed. Retroviruses enter the cell through the cell wall and implant their DNA into the cell's nucleus. The retroviral DNA is incorporated into the cell's DNA, causing the disease that the particular retrovirus is associated with. Scientists reasoned that if they could put a gene into the retrovirus, the retrovirus would deliver that gene to the cell's DNA. An added bonus is that different retroviruses target different areas of the body, so the scientists could put DNA into a retrovirus for delivery to a specific organ.

Attaching the New DNA

Once the cell's original DNA is cut with the restriction enzyme and the new DNA is in place, the scientists use ligase (another enzyme) to stick the DNA segments together. The sticky ends of the new and the original DNA merge together, and the cell begins to carry out the instructions of the new DNA along with its own.

Atrition

Of course, this technique does not work on 100% of the cells. A scientist may start with a lot of cells, and the restriction enzyme may not get to some. The DNA may not enter some of the cells. Other cells' DNA will not line up properly with the new DNA, causing the ligase to bond the DNA incorrectly.

The new DNA may find its way to the wrong position along the original DNA. In others, the ligase may not get to the cell and fuse the DNA at all. In every "batch" of genetically altered cells, there will be some that do not have the new trait. The cells are grown in culture, and the cells that do not posses the desired trait are culled out, usually by subjecting all the cells to whatever treatment the new cells were designed to withstand. The cells that survive have the new DNA properly aligned, and are ready to go to work.

Safeguards on Genetic Engineering

Sci-fi writers often describe worlds where genetic engineering has gotten out of hand, and has destroyed some aspect of society. But how likely is this to really happen?

Misconceptions

Saturday morning cartoons capitalize on the notion that scientists plunge into the unknown without concern for the consequences. They don't. A scientist is not, in all probability, going to create a supervirus or a man-eating plant just to see if he can. Scientists realize the possible effects of all their research, and stop if there's a possibility of danger.

Precautions

Experience has shown that when one aspect of nature is changed, there can be unexpected repercussions in other areas of the ecosystem or in the economy. Because of this, scientists must be very careful about making sure that any new organism that they create will not damage the life that already exists.

Let us use the example of genetically altered fish. After much labour, scientists have succeeded in creating a fish that grows up to 40% faster than the average fish. The goal in this research was to make fish for human consumption that grow faster, and are ready for harvest sooner.

The first step toward assuring that the fish are viable and safe is to put them in a microcosm, an enclosed, controlled replica of the ecosystem in which they were designed to live. During this time, the scientists monitor the fish to see if the implanted gene is working properly. They are also looking to answer several questions that will determine if the fish are safe to release into the wild.

- Will the organism survive? If the fish can't survive in the wild, it goes "back to the drawing board". Fortunately, this new fish is capable of living on its own.

- Does the organism reproduce, and if so, how fast? If it multiplies rampantly, it may be a disastrous addition to the ecosystem, upsetting the delicate balance. If it does not multiply at all, it still passes this test, but it may be advantageous to allow the organism to multiply. Scientists find that the fish is capable of reproduction, and at a normal rate.

- Does the organism have an advantage over other organisms, and if so, does it compete for the same resources? Imagine the catastrophe if an organism was so strong and took so many natural resources out of the environment that it killed off all its neighbours! Scientists need to find out if this organism will lead to the extinction of others. Unfortunately, this fish competes for the same resources as normal fish, and wins. Because it is growing so rapidly, it needs a lot of food. It would probably eat the other fish out of a habitat. This, coupled with its size, could lead to the destruction of the ecosystem.

- Can the gene move to other species? There is no way to predict what will happen when a fish with an altered gene is bred with a normal fish. Just in case the gene would be harmful in other organisms, the scientists need to make sure that the gene cannot be passed along. Unfortunately, scientists find that the growth gene in the altered fish can be passed on to other fish.

- Does the organism leave the test area? This may sound scary at first, but basically scientists want to know whether the organism will move itself to anywhere habitable or whether it will mostly stay where it is released. In the case of our fish, their ravenous appetite almost forces them to move on in search of more food.

The genetically altered, fast-growing fish have failed three tests. The scientists must change the fish so that they pass all the tests, or they cannot be released. In order to pass the fourth test (can the gene move across species), a third chromosome is inserted.

This third chromosome keeps the fish from being able to form gametes with the correct number of chromosomes. Offspring cannot be conceived, so the fish do not reproduce at all. No children means no gene passing. In order to pass the fifth test, the fish are enclosed under three layers of fencing, one of which is electrified. With the fish placed in this controlled environment, the third test is automatically passed; there are no other fish with which to compete.

Ripples

There are other implications from genetic engineering besides danger to health. For instance, the plant that usually gives us canola oil can be modified to produce lauric oils, which were once only produceable in Southeast Asia. While it is cheaper and more efficient to grow the new plant than to buy the old, American money is not going into Southeast Asia to buy the oil. The Southeast Asian economy has been effected by the new plant.

Fruit is being genetically engineered to stay fresh longer. However, many suspect that this is a counterfeit freshness; the vitamins and minerals that are the reason to eat fruit at all are gone long before the fruit begins to go bad. This has made the fruit about as healthy as a regular, processed food, even though it appears very ripe.

Some farmers use a crop with a built-in insecticide that kills pests. As always, there are certain bugs that are resistant to the insecticide. The insecticide gives the mutant bugs a competitive advantage, and there is a threat that they will multiply rampantly with the farmers providing their food source.

Another insecticide must be used on these bugs, which is dangerous to helpful insects as well as the crop. Farmers have had to plant standard crops next to the engineered ones so that the original bugs are still being provided with food and can multiply, keeping the resistant strain of insect from having and advantage.

Even though new organisms don't appear to hurt, the lesson is that they may not help, and may even be causing damage in places that were totally unpredictable. Great care must be taken when altering the tried-and-true genetic code that already exists.

Ethics of Genetic Engineering

Genetic engineering offers a major ethical dilemma — there are great rewards if genetic engineering becomes a widespread reality, but the dangers are equally great. In addition, genetic engineering also offers the ability to change the very nature of nature, an environment in which man is already well-suited to live in. Is it possible that genetic engineering will do more harm than good?

Medicines

As you have probably already learned, DNA is very versatile because all life on Earth processes it the same way (as far as we know). Therefore, foreign genes can be implanted in bacteria, instructing the bacteria to produce whatever the inserted gene controls.

In this manner, scientists can "teach" bacteria to produce human or other proteins. There is a long list of bacteria that have been treated this way, including an e. coli bacteria that produces insulin. The insulin is less expensive than that removed from cows and pigs, as well as being exactly the same as human insulin.

The ethical quandary about this process lies in the fact that human genes are being inserted into a bacteria. I do not know of anyone who stands up for bacterial rights; in fact, everyone who has been sick is rather prejudiced against bacteria.

The problem is the idea that human genes are being put in a non-human organism. Is there anything special about human genes? If you consider that human genes are made of the same four bases as bacterial genes, and that they work the same way — as an instruction manual for the production of a protein, there is less of a dilemma.

However, many people believe that human genes should not be implanted in other species, because they are special because they are human. Before you make up your mind, consider the alternative method of production, and consider honestly whether or not you would be willing to recieve medicine from a human gene in a bacteria if it was you who needed it.

Xenotransplantation

Work has been done on genetically engineering farm animals, too. While in the future it may provide leaner meat, larger animals, or faster growth, right now the great prospect is xenotransplantation. Xenotransplantation transplanting organs across species. Currently, most organ transplants are human to human. They work well, but the drawback is the rejection medication that must be taken to weaken the immune system so that the body does not destroy the new organ. This, of course, can lead to illness, a definite drawback. If organs can be "made to order" in another organism and then transplanted over to the human patient, this can eliminate the need for rejection medication and the succeptibility to illness.

The best prospect for xenotransplantaton of vital organs is the pig. The pig is one of the closest genetic matches to humans, closer even than monkeys and apes. With a little genetic alteration, pigs can hopefully produce organs that the human immune system will not reject, producing the desired effect. Unfortunately, the pig is the loser in this arrangement.

While on the surface this may seem barbarous, consider the concept: a human is taking something from the animal to extend his own life. The basic principle is not all that different from eating. If xenotransplantation still seems cruel, perhaps the world needs to re-evaluate itself and turn vegetarian, eliminating the need to raise animals for the sole purpose of death for man's benefit.

Custom-made Organisms

Current technology and a great deal of experimentation may allow for custom-made organisms to be produced. These may range from a bacteria that cleans up oil spills to a monkey with gills and a higher intelligence level. Everyone seems to agree that we should not pursue these advances just for the fun of it, so the question becomes, are we so inconvenienced by a problem that we need to create a new life form to fix it for us?

Core Issue

The core issue behind the ethics of genetic engineering is as follows. Is it morally right to change the nature of life on Earth to suit man's desires better? The answer to this question depends on man's position on Earth. If we are truly superior to the animals and accountable to no one, is there a real question of whether it is wrong? If we are not fundamentally different, do we have a right to meddle with evolution? If we are accountable to God for our actions, should we risk insulting His creation by trying to do it better?

Future of Genetic Engineering

Genetic Engineering is, quite possibly, the most promising and most threatening new advance in medical technology of all time. For this reason, new advances made in the future may or may not be implemented. It is nearly impossible to predict what Genetic Engineering will bring to the future, but there is some speculation.

Medicine

The most popular contribution of Genetic Engineering is in the field of medicine. Because most diseases have a genetic component, healing may be sped along in the future by using bacteria-made proteins to augment the healing process. Proteins circulate through the body, reporting to various glands whether bone and muscle mass should increase or decrease, how salty substances should be moved through the lungs, how many immune cells should be manufactured, and even how fast hair should grow. The genes regulating the production of these proteins can be inserted into bacteria, "teaching" the bacteria how to make the identical proteins.

Unfortunately, this is a power that is easily abused. Athletes could have their muscle mass increased for more strength. It could be that the army with the most genetic drugs coursing through their veins will be the one to win the war. Many people feel that these dangers far outweigh the good that could be done speeding the healing process.

Food

One controversial advance has been in place for several years: genetically engineered food. Dubbed "Frankenfoods" by opponents, these foods are enhanced by genes from other plants or animals. This can enhance nutrition, shelf life, or taste. Depending on public reaction, these foods may be widely available in the future.

Genetic Mapping

Find out about the Human Genome Project and other genome mapping projects.

The Human Genome project and other genetic mapping endeavours hope to discover what makes man "tick". So doing, they may be able to help doctors test succeptibility to disease and allergies to drugs. They may be able to help genetic engineers by giving them more insight into where traits are located. Many areas of science will be affected by genetic mapping. Read on to learn how scientists are going about this herculean effort.

Overview of Genetic Mapping

What is the Human Genome Project all about?

As you probably know from reading the Textbook section, DNA is read by the cell in triplets. Each set of three base pairs codes for an amino acid, which is the building block for the over 50,000 kinds of proteins in your body. These proteins direct growth and allow your body to function properly. The Human Genome Project hopes to find out what each segment of your DNA does.

The Basics

The Human Genome Project will cost nearly 40% the money required to build the International Space Station. It is funded by the United States Department of Energy and the National Institutes of Health. The 15 year project is working with over 400 teams, although half the budget goes to 21 main research centres.

The Human Genome Project is scheduled to be finished in 2005. Its basic goal is to find all the genes in the human body, as well as other model organisms, and to determine the function of those genes. However, with 1990's technology, that goal was not accomplishable in 15 years. Therefore, the goals of the project also include the development of new technology to rapidly sequence and interpret the DNA.

History

Back in 1985, a group of physicians got together and decided that the medical community should declare war on cancer. Of course, the first order of business was to figure out what caused cancer in the first place. Well, radiation can cause cancer. Why does radiation cause cancer? Because it alters the genetic structure of the cell. What part of the DNA is altered? No one knew. At that moment, the idea of mapping the human genome became very popular. In 1990, the Department of Energy and the National Institutes of Health launched the Human Genome Project.

Goals

In addition to mapping the human genome, the project is also mapping several model organisms — animals that share certain aspects of humanity. Among a few others, scientists are mapping the mouse (vertebrate; very similar to humans), the roundworm (multicellular), a yeast (cell with nucleus), and E. coli (alive). They chose these organism because they each share progressively less in common with humans, and they have all been subjects of scientific scrutiny for some time. In fact, most genetic engineering work is done either with mice or E. coli bacteria. Please note that the scientists are working with a harmless form of E. coli and not its dangerous cousin.

The Human Genome Project's goal, basically stated, is to map the entire human genome, find out what each of your genes does, and discover the sequence of the base pairs of those genes. The data gathered in this project can be used to develop tests for genetic conditions and susceptibilities.

This information is potentially very useful; if a man finds that he is genetically prone to a heart attack, he may eat a healthier diet or begin to exercise. The same testing can also be used to predict side effects or effectiveness of medication. It can determine if you are a carrier for certain recessive diseases. The vast amount of data collected by the Human Genome Project can be put to very good use, but it can also be abused; discrimination based on "faulty" genes may be a very real issue in the future. Therefore, the Human Genome Project also devotes 5% of its budget to looking at the implications of its research.

Ethics

One of the unique aspects of the Human Genome Project is that it has a sub-program called ELSI that looks at the Ethical, Legal, and Social Implications of the Human Genome Project, trying to address issues before they come up.

Genetic testing holds bright promise for the near future. However, there is also a great debate about what sort of information should be made available, and to whom. For instance, who is allowed access to genetic information? You? Your doctor? Adoption agencies? Employers? Insurance companies? The government? Access isn't the only problem. If you found out that you would become terminally ill at age 45, and there was nothing you could do about it, would that affect your life so adversely that it would be better not to know? ELSI tries to come up with answers to tough questions before all of the Human Genome Project's data is acquired and released.

Process of Genetic Mapping

Learn how scientists go from AACTGG... to an understanding of how a particular gene works.

Human DNA is astoundingly complicated. The four bases of DNA make up the six feet of genetic material in every cell. This material is spread out over 23 chromosomes. The three billion base pairs in a single human cell make up 60,000-70,000 genes, plus the introns (about 95%) whose function still remains a mystery. It can be plainly seen that constructing a genetic map is a daunting task.

Getting Started

Many people want to know exactly whose DNA geneticists are mapping. The Human Genome Project is not trying to map a specific person's DNA, but DNA in general. 99.9% of the DNA of every person on earth is identical. Scattered throughout this identical DNA is the 0.1% (3,000,000) of base pairs that are vary from person to person: the variations and mutations that make one human different from another. These variations provide the key to unlocking the human genome. By comparing the DNA of many different people, especially those with genetic disorders, scientists hope that they can map the human genome by 2005.

Genetic Linkage Maps

The first step toward mapping the genome is to establish where one gene is in relation to another. This is done by studying people with genetic disorders. During meiosis, chromosomes pair up and exchange parts of their DNA, mixing the parents' chromosomes into a single chromosome that will be passed on to the offspring in the form of a sperm or an egg cell. This process is called "crossing over".

Two traits that are often inherited with each other are almost definitely very physically close to each other on the chromosome, because the closer two genes are to each other, the more likely they are to move together when the chromosomes mix.

Genes that often move together when chromosomes cross over are said to be linked. Although a genetic linkage map does not show a gene's location, it can show the odds that one specific gene will be inherited with another specific gene; in fact, the distance between genes on genetic linkage maps is not measured in terms of physical distance, but in terms of what the probability is that two genes will cross over together.

Contig Maps

Once scientists determine which genes are linked together, they can begin to form a contig(uous) map of the chromosome. A contig map is designed to show the general physical location of specific genes on a chromosome; the distance between genes on a contig map is measured in terms of physical distance, as opposed to the genetic linkage map, which measures the probability of two genes being inherited together. The margin of error when measuring the distance between two genes on a contig map is 500,000 base pairs. A contig map can also show which blocks of chromosomes are mostly coding genes and which blocks are mostly introns.

Genetic Sequencing

The ultimate goal of the Human Genome Project is to produce a genetic sequence map of the human genome. At the project's conception in 1990, computer technology was not good enough to map the entire genome in anywhere close to 15 years. As computing technology improves and scientists' technique is revised, the dream of a genetic sequence is drawing closer and closer to being realized. Of course, with genetic sequencing, the problem again arises, "Whose DNA is going to be mapped?" When the scientists reach the areas where individuals' DNA differs, all the participants' individual DNA will be sequenced and catalogued.

What Next?

Even after the Human Genome Project is finished, there will be the task of interpreting all the data. Scientists still need to learn about the function of introns, about the different possible combinations of

DNA that yield a specific trait, and about the effects of mutations. The Human Genome Project in no way marks the end of discovery in the human genome.

Future of Genetic Mapping

Find out what Genetic Mapping may bring us in the Future

Genetic mapping is a project that has been undertaken by the American government in conjunction with private labs. How can a project have a future if it has a logical ending point? While the project itself will be complete, the information gleaned and the technology developed will be in use long after the project is complete.

Medicine

Many scientists see the first mainstream applications of genetic mapping to be in the field of medicine, specifically, predicting disease susceptibilities and drug allergies. Each patient's genetic code is different, each with different strengths and weaknesses. If a genetic map reveals to a doctor that a specific disease is likely, the doctor and patient can attempt to prevent the disease — susceptibility to heart disease can be countered by a healthy diet, for instance.

A young man was recovering from surgery for an ear infection. He was taking antibiotics, but he began to feel very bad, and the doctors feared that another surgery would be necessary. However, his friend recognized his symptoms and reported that one of her relatives had the same symptoms, which were eventually traced to a rare allergy to the antibiotic that she was taking.

The young man was taking the same antibiotic. When the drug was replaced, he began to feel better and eventually recover. He later learned that he might have died from continued use of the antibiotic. If genetic testing were available, it could have prevented his problems. This sort of genetic test may be available in the relatively near future.

Identification

The Human Genome project is not only developing a map of the genome, but also new technology to speed sequencing and analysis. If sequencing technology continues to develop at its current rate, it may be possible in the future to perform a genetic identity test in minutes; some predict that this test may even be performed with a hand-held device. Of course, this would require a database with portions of everyone's genetic code. This advance may be prevented if people refuse to be catalogued.

Genetic Engineering

Genetic sequence maps will show what sequences of DNA create what proteins and what those proteins do. This knowledge, when catalogued, will allow genetic engineers to find and insert combinations of DNA sequences into organisms, giving the new organisms desirable traits. Genetic maps will allow these scientists to predict more accurately how traits will interact and allow for more radical changes of the organisms.

DNA Fingerprinting

This area is about how scientists can use DNA to positively identify a person or his involvement in a crime.

Many people know about DNA Fingerprinting, but very few know how the process works. Here, you will learn what scientists must do to determine if a person and a DNA sample match.

Fingerprinting a Crime Suspect

There are three basic components to a DNA fingerprint test. First, there is the DNA from the crime scene; second, there is the DNA from a suspect; third, there is DNA with known attributes that is used as a control. All the DNA runs through the same tests at the same time. First, the three sets of DNA are purified and replicated, giving the scientists more DNA to work with. Next, the DNA is cut using a restriction enzyme. All three sets of DNA are then placed on the same block of gel.

When electricity is passed through the gel, the DNA will migrate down the gel block; short strands will move farther down the gel than the long strands. This organizes the DNA strands by length so that the results can be accurately measured.

A thin film of nylon is laid over the gel to blot up the DNA. Finally, short, radioactive strands of DNA called probes are sent onto the nylon. These radioactive probes will pair up with complementing strands of DNA on the nylon. The probes are designed to seek out and pair up with uncommon sequences of DNA; the probes will all congregate to the places where the uncommon sequences from the replicated DNA have come to rest.

When the nylon film is X-rayed, any large concentrations of the radioactive probes will leave blots on the X-ray print. If the blots from the control DNA look different than they normally do, the scientists know that there was error in the test, and the results are thrown out.

If the control blots are in the correct positions, the test was run properly. If the blots from the suspect line up with the blots from the DNA gathered at the crime scene, then the two samples came from the same person, and the suspect was indeed present at the crime scene.

Error

Many have questioned the validity of DNA fingerprinting test results. Because the burden of proof is on the forensic scientists, there are safeguards in place to be sure that DNA fingerprinting results are accurate.

The first is the control DNA that runs through the tests alongside the suspect's DNA and the DNA found at the crime scene. If the control DNA tests properly, then the DNA in question tested properly as well.

Another safeguard in place is that the test is run four times using four different restriction enzymes. In order to be absolutely positive of a match, all four sets of DNA from the crime scene must match all four sets of DNA from the suspect. With these safeguards in place, the odds are one in 70 billion (140 times Earth's population) that the results are inaccurate.

However, DNA fingerprinting is sometimes refutable evidence, even with such low odds of innacuracy. If the technician confused the DNA samples or deliberately gave false results, the information gleaned from a DNA test would be useless.

Also, DNA evidence only proves that a person was present. It does not prove that the depositor of the DNA was the actual criminal. For these reasons, there must be other evidence to support the DNA fingerprints' conclusion; the fingerprints alone are not enough for a conviction.

Applications of DNA Profiling

Nearly everyone knows that DNA Fingerprinting can be used to prove a person's connection to a crime. Few realize the other applications of DNA fingerprinting.

Establishing Innocence

DNA Fingerprinting can help to speed along court cases by eliminating suspects. Between 1989 and 1996, the FBI used genetic testing in about 10,000 sexual assault cases; in 2,000 of those cases, the prime suspect was discovered to have not committed the crime. Without genetic testing, it can be assumed that some of these men would have been convicted. In fact, many prison inmates have appealed their conviction after spending years in jail, and have been discovered to be innocent.

Identification

As a result of the unidentified remains of American soldiers that have been found abroad, the U.S. military has been compiling data from mitochondrial DNA to help identify the soldiers. Mitochondrial DNA is passed on by the mother only, and as a result, remains mostly identical within families. By testing DNA from the bones of the remains of soldiers against the military's database, scientists have identified 500 soldiers so far, and another 250 more are under study.

Ethics

There many other genetic tests available now that can determine identity and paternity. When the Human Genome Project is completed, scientists predict that a battery of tests will be available to determine what diseases and characteristics a person is prone to. But as with any new knowledge, we must determine what people have access to what information, and how that information can be used.

Identity Test

Currently, DNA testing can be used to match one genetic sample to another. This is being used in many court cases to prove that a person was at a scene where his DNA was picked up in the form of a hair, blood, or skin samples. However, the police cannot just look at some DNA and know whose it is; they must have a suspect, get a DNA sample from him, and attempt a match. Without a suspect, there is no way to figure out whose DNA was at the scene.

Many people believe that crimes could be solved much more quickly and efficiently if there was a national database containing DNA samples from each person. Of course, the entire DNA strand would not be needed — just the pieces usually used for testing. But should the government really have access to everyone's DNA samples?

Full Workup

Scientists hope that in the near future, testing will be available that will determine immunity or succeptibility to disease, whether the patient will have side effects to certain drugs, whether he might develop a terminal illness, or even what his basic intelligence potential is. That's a lot of information, and a good bit of it could be damaging in the wrong hands. But whose hands are the wrong hands? This full genetic workup leads to many, many questions. For instance, should this test be done at birth or only on demand? Should insurance companies allowed to see your disease succeptibilities? Can they charge you more if you are succeptible to an expensive disorder? Should drugs be allowed on the market that can be very harmful to some but not to others, provided that the recipient is tested to make sure the drug is safe for him?

If a person will likely develop a terminal illness, should the doctor tell him, or just decide that ignorance is bliss and let him find out twenty, thirty, even fifty years down the road? Can an employer discriminate based on basic intelligence? These questions only scratch the surface of what may come. In addition, if access to a product is restricted, a black market usually springs up. What should the penalty be for trying to gain access to another person's genetic information?

Core Issue

The core issue brought up by genetic testing is this — is a person bound to a genetic future, or is man more than the outcome of his genes? Is genetic tendency immutable? Can a little strand of chemicals really determine the outcome of our entire lives?

Eugenics

The concept of moving selective breeding from the animal world to the human world began to take root in the 1830's. Could this natural extension of Genetics and Darwinism be resurfacing?

Roots

In 1883, Francis Galton, cousin of Charles Darwin, coined the word "Eugenics", which means "good in birth". In his book *Natural Inheritance* (1889), he theorized that society would be better without the presence of inferior people. In fact, he continued this line of reasoning to say that societies topple because the genetically superior aristocracy have fewer children than the inferior peasants. To keep this from happening, he proposed two methods of control — Positive eugenics would create incentives for the very intelligent or successful to reproduce. Negative eugenics would keep the unfit from reproducing.

Applications

Galton's ideas grew in popularity both in America and abroad. In 1896, Conneticut had put in place laws that forbade sex with epileptics or the mentally handicapped. In 1907, a state law was passed that required sterilization of institutionalized, mentally retarded males, as well as criminal males.

This law would eventually lead to other states' passage of similar laws; before they were repealed in 1956, 58,000 American citizens were sterilized. The Immigration Act of 1924 limited the immigration of "idiots, imbeciles, feebleminded, epileptics, insane persons" to the United States.

The philosophy of eugenics had taken root in Nazi Germany, as well. In 1934, the Eugenic Sterilization Law was passed in Nazi Germany, ordering the sterilization of individuals with genetically inherited diseases; the Genetic Health Courts would decide who would and who would not be sterilized. In 1939, the situation turned very ugly. The Nazi laws were extended to allow the killing of genetically undesirable people; by the time World War II was over, 75,000 "unfit" persons had been killed, including 5,000 children. As the world discovered the atrocities tha had been committed, many began to rethink their positions regarding these laws. In 1956, all United States eugenic sterilization laws were repealed, and nine years later, many of the immigration restrictions were lifted.

Modern Eugenics?

Thanks to the Human Genome Project and other genetic research, prenatal testing allows for the detection of many illnesses, including cystic fibrosis and Down's syndrome. Pregnant women can now know ahead of time if their fetus will contract an incurable disease or condition.

Often, a diagnosis leads to the termination of the fetus. This can be construed to be a combination of euthanasia and eugenics, an attempt to keep "bad genes" out of the gene pool, or as an act of mercy, terminating a fetus whose quality of life is in doubt. But which is it? And whose decision is this really to make?

Who gets to decide what diseases/conditions are too bad to have to live with and what diseases/conditions are not so bad?

Ethics of Genetic Engineering

Genetic engineering offers a major ethical dilemma — there are great rewards if genetic engineering becomes a widespread reality, but the dangers are equally great. In addition, genetic engineering also offers the ability to change the very nature of nature, an environment in which man is already well-suited to live in. Is it possible that genetic engineering will do more harm than good?

Medicines

As you have probably already learned, DNA is very versatile because all life on Earth processes it the same way (as far as we know). Therefore, foreign genes can be implanted in bacteria, instructing the bacteria to produce whatever the inserted gene controls. In this manner, scientists can "teach" bacteria to produce human or other proteins. There is a long list of bacteria that have been treated this way, including an e. coli bacteria that produces insulin. The insulin is less expensive than that removed from cows and pigs, as well as being exactly the same as human insulin. The ethical quandary about this process lies in the fact that human genes are being inserted into a bacteria. I do not know of anyone who stands up for bacterial rights; in fact, everyone who has been sick is rather prejudiced against bacteria. The problem is the idea that human genes are being put in a non-human organism. Is there anything special about human genes? If you consider that human genes are made of the same four bases as bacterial genes, and that they work the same way — as an instruction manual for the production of a protein, there is less of a dilemma.

However, many people believe that human genes should not be implanted in other species, because they are special because they are human. Before you make up your mind, consider the alternative method of production, and consider honestly whether or not you would be willing to recieve medicine from a human gene in a bacteria if it was you who needed it.

Xenotransplantation

Work has been done on genetically engineering farm animals, too. While in the future it may provide leaner meat, larger animals, or faster growth, right now the great prospect is xenotransplantation. Xenotransplantation transplanting organs across species. Currently, most organ transplants are human to human. They work well, but the drawback is the rejection medication that must be taken to weaken the immune system so that the body does not destroy the new organ. This, of course, can lead to illness, a definite drawback. If organs can be "made to order" in another organism and then transplanted over to the human patient, this can eliminate the need for rejection medication and the succeptibility to illness.

The best prospect for xenotransplantaton of vital organs is the pig. The pig is one of the closest genetic matches to humans, closer even than monkeys and apes. With a little genetic alteration, pigs can hopefully produce organs that the human immune system will not reject, producing the desired effect. Unfortunately, the pig is the loser in this arrangement. While on the surface this may seem barbarous, consider the concept: a human is taking something from the animal to extend his own life. The basic principle is not all that different from eating. If xenotransplantation still seems cruel, perhaps the world needs to re-evaluate itself and turn vegetarian, eliminating the need to raise animals for the sole purpose of death for man's benefit.

Custom-made Organisms

Current technology and a great deal of experimentation may allow for custom-made organisms to be produced. These may range from a bacteria that cleans up oil spills to a monkey with gills and a higher intelligence level. Everyone seems to agree that we should not pursue these advances just for the fun of it, so the question becomes, are we so inconvenienced by a problem that we need to create a new life form to fix it for us?

Core Issue

The core issue behind the ethics of genetic engineering is as follows. Is it morally right to change the nature of life on Earth to suit man's desires better? The answer to this question depends on man's position on Earth. If we are truly superior to the animals and accountable to no one, is there a real question of whether it is wrong? If we are not fundamentally different, do we have a right to meddle with evolution? If we are accountable to God for our actions, should we risk insulting His creation by trying to do it better?

Current Developments

Transgenics allow scientists to develop organisms that express a novel trait not normally found in the species; for example, a type of rice known as golden rice has elevated levels of vitamin A. Scientists have also developed sunflowers that are resistant to mildew and cotton that resists insect damage. Possible transgenic combinations can be broken down generally into three categories (here "animal" refers to nonhumans):

- plant-animal-human combinations
- animal-animal combinations
- animal-human combinations

An example of a plant-animal-human transgenic combination would be one in which the DNA of mouse and human tumor fragments is inserted into tobacco DNA. The harvested plants contain a potential vaccine against non-Hodgkin's lymphoma. Other transgenic plants have been used to create edible vaccines. By incorporating a human protein into bananas, potatoes, and tomatoes, researchers have been able to create prototypes of edible vaccines against hepatitis B, cholera, and diarrhea. The vaccines are proving to be successful in tests on agricultural animals and humans.

A product created from an animal-animal transgenic combination. Scientists at Nexia Biotechnologies, a company based in Montreal, isolated the gene for silk protein from a spider capable of spinning silk fibers—one of the strongest yet most resilient substances known—and inserted it in the genome of a goat's egg prior to fertilization. When the transgenic female goats matured, they produced milk containing the protein from which spider silk is made. The fibre artificially created from this silk protein has several potentially valuable uses, such as making lightweight, strong, yet supple bulletproof vests. Other industrial and medical applications include stronger automotive and aerospace components and stronger, more biodegradable sutures for closing wounds.

Bibliography

Alfred Steferud: *Diseases of Fruits and Nuts*, Biotech Books, Delhi, 2005.

Arthur W. Gilbert, Mortier F. Barrus and Daniel Dean: *Growing and Breeding of Potatoes*, Asiatic Pub, Delhi, 2006.

Ashworth S.: *Seed to Seed*, Decorah, Seed Savers Publications, 1991.

Ausubel, F.M. : *Current Protocols in Molecular Biology*, New York: John Wiley and Sons, 1989.

Bahar A. Siddiqui and Samiullah Khan: *Plant Breeding Advances and in vitro Culture*, CBS, Delhi, 1997.

Banki, L.: *Bioassay of Pesticides in the Laboratory*, Akademiai Kiado, Budapest, 1978.

Barbeau, G.: *Tropical Fruits in Nicaragua*, Managua, Nicaragua Ministerio de Desarrollo Agropecuario, Agraria, 1990.

Barnes, N.: *Biology*, New York, Worth Publishers, 1989.

Barnum, Susan R.: *Biotechnology: An Introduction*, Belmont, Thomson/Brooks/Cole, 2005.

Barrington, E. J. W.: *Biochemistry of Primitive Deuterostomians*, London, Academic Press, 1974.

Baruah, Akhil: *Advanced Morphology of Angiosperms*, Aavishkar, Delhi, 2008.

Batlle I. and Tous J.: *Carob Tree (Ceratonia siliqua L.)*. Rome, International Plant Genetic Resources Institute, 1997.

Bauer MW: *Biotechnology-the Making of a Global Controversy*, Cambridge, Cambridge University Press, 2002.

Bhatia, A L : *Biochemistry and Endocrinology,* Indus Valley, Delhi, 2002.

Bhatnagar, Vasudev : *Cell Science and Technology*, Campus Books, Delhi, 2009.

Brandwein, P.F. : *Sourcebook for the Biological Sciences,* San Diego: Harcourt Brace Jovanovich, 1986.

Broach, J.R.: *The Molecular Biology of the Yeast*, Cold Spring Harbor, Cold Spring Harbor Laboratory, 1981.

Brodwin, Paul : *Biotechnology and Culture: Bodies, Anxieties, Ethics*, Bloomington: Indiana University Press, 2001.

Cahill, Lisa : *Genetics, Theology, and Ethics: An Interdisciplinary Conversation*, New York: Crossroad, 2005.

Chaudhary, Vikas: *Entomology and Pest Management*, Navyug, Delhi, 2008.

Chauhan, B.S. : *Principles of Biochemistry and Biophysics*, Laxmi Publications, Delhi, 2008.

Chiranjib Chakraborty: *Advances in Biochemistry and Biotechnology*, Daya, Delhi, 2005.

Chrispeels, Maarten : *Plants, Genes and Crop Biotechnology*, Sudbury MA, Jones and Barlett Publishers, 2003.

Clark, J.M.: *Experimental Biochemistry*, New York, W.H. Freeman and Company, 1977.

Clawson, Calvin: *The Mathematical Traveller*, New York, Plenum Press, 1994.

Collins, Steven : *The Race to Commercialize Biotechnology: Molecules, Market and the State in Japan and the US.*, New York: Routledge, 2004.

Collymore L.: *Fruit Production in Barbados*, Port of Spain, Trinidad and Tobago, 1996.

Coste R.: *Coffee: the Plant and the Product*, London, MacMillan, 1992.

Cronquist, A.: *The Evolution and Classification of Flowering Plants*, New York Botanical Garden, Bronx, New York., 1988

Currah L. and Proctor F. J.: *Onions in Tropical Regions*, Kent, Natural Resources Institute, 1990.

Dabholkar, A.R.: *General Plant Breeding*, Concept, Delhi, 2006.

Daphne C. Elliott: *Biochemistry and Molecular Biology*, Oxford University Press, Delhi, 2005.

David Sadava: *Plants, Genes and Crop Biotechnology*, Sudbury MA, Jones and Barlett Publishers, 2003.

Dixon, Dougal: *After Man-A Zoology of the Future*, New York, St. Martin's Press, 1981.

Dodds, John H.: *Plant Genetic Engineering*, New York, Cambridge University Press, 1985.

Doijode S. D.: *Seed Germination in Fruits*, New Delhi, Malhotra Publishers, 1993.

Duckworth, W. L. H.: *Morphology and Anthropology : A Handbook for Students*, Cosmo, Delhi, 2006.

Dudley, E.: *The Critical Villager: Beyond Community Participation*, London, Routledge, 1993.

E. Ramann: *The Evolution and Classification of Soils*, Asiatic Pub, Calcutta, 2006.

Elliott, B N : *Biochemistry and Molecular Biology*, Oxford University Press, Delhi, 2005.

Featherly H. I.: *Taxonomic Terminology of the Higher Plants*, USA, Iowa State College Press, 1954.

Ferentinos L.: *Proceeding of the Sustainable Taro Culture for the Pacific Conference*, Honolulu, HITAHR, 1993.

Fransman M, Junne G, Roobeek A: *The Biotechnology Revolution?*, Oxford, Blackwell, 1995.

Friedberg, E.C.: *DNA Repair*, New York, WH Freeman and Company, 1985.

Fumento, Michael: *Bioevolution: How Biotechnology is Changing Our World*, San Francisco, Encounter Books, 2003.

Ganguly, Smriti : *Biochemistry of Biomolecules,* Pearl Books, Delhi, 2007.

Ghulam Hassan : *Soil Microbiology and Biochemistry,* New India Publishing Agency, Delhi, 2010.

Goodsell, David S.: *Bionanotechnology: Lessons From Nature*, Hoboken, Wiley-Liss, 2004.

Graf, Alfred Byrd: *Advances in Plant Physiology*, Rajat Pub, Delhi, 2008.

Hardy B.: *Biology and Agronomy of Forage Arachis*, Cali, International Centre for Tropical Agriculture, 1994.

Jeffers P.: *Evaluation of Four Onion Varieties in Montserrat*, Plymouth, CARDI, 1992.

Jones, R. M.: *Plant Resources of South-East Asia,* Wageningen, Pudoc Scientific Publishers, 1992.

Khanna, V K : *Objective Genetics, Biotechnology, Biochemistry and Forestry,* I.K. International Publishing House, Delhi, 2008.

Kuppuram, G & K. Kumudamani: *History of Science and Technology in India*, Delhi, Sundeep Prakashan, 1990.

Kurzweil, Ray: *The Age of Spiritual Machines*, New York, Penguin Books, 1999.

Larry V. McIntire : *Biotechnology: Science, Engineering, and Ethical Challenges for the Twenty-first Century,* Washington, DC: Joseph Henry Press, 1996.

Macself, A.J.: *Soils and Fertilizers,* Satish Serial Pub, Delhi, 2005.

Madan Lal Bagdi: *Physiology, Biochemistry and Biotechnology*, Manglam Pub, Delhi, 2007.

Madulid Domingo A.: *A Pictorial Cyclopedia of Philippine Ornamental Plants*, Philippines, Makati Metro Manila, 1995.

Mahindru, S. N.: *Food Safety and Pesticides*, APH, Delhi, 2009.

Meena Francis: *Biotech's Dictionary of Biochemistry*, Biotech Books, Delhi, 2007.

Muneesh Kainth: *Chordate Embryology*, Dominant, Delhi, 2003.

Nobel, P. S.: *Physicochemical and Environmental Plant Physiology*, Academic Press, San Diego, 1999.

Old, R.W. : *Principles of Gene Manipulation*, London, Blackwell Scientific Publications, 1989.

Oldham P.: *Cost of Production of Major Tree Crops in Dominica*, Roseau, Ministry of Agriculture, 1991.

Parry M.L.: *Climatic Change, Agriculture and Settlements*, Dawson Folkestone UK, 1978.

Paul M. Althouse: *Introduction to Agricultural Biochemistry*, Biotech Books, Delhi, 2005.

Pemberton, R. W.: *Predictable Risk to Native Plants in Weed Biological Control,* Oecologia, 2000.

Qystein V. Sjaastad: *Physiology of Domestic Animals*, International Book Distributing Co., Delhi, 2005.

Ragone D.: *Breadfruit: Artocarpus Altilis (Parkinson) Fosberg*, Rome, International Plant Genetic Resources Institute, 1997.

Rifkin, Jeremy: *The Biotech Century*, New York, Penguin Putnam, 1998

Rutherford Lyn.: *A Gourmet's Book of Mushrooms & Truffles*, Sydney, Golden Press Pvt. Ltd., 1991.

Sharma, Pradeep : *Biochemistry and Organisation of Cells*, RBSA Pub, Delhi, 2006.

Stover, R. H. and Simmonds N. W.: *Bananas*, United Kingdom: Longman Scientific and Technical, 1991.

Swarnim, K. : *A Textbook of Biochemistry and Microbiology*, Surendra Pub, Delhi, 2010.

Tawde, A. B.: *Propagation and Rootstocks of Mango*, New Delhi, Malhotra, 1993.

Urton, Gary: *The Social Life of Numbers*, Austin, University of Texas Press, 1997.

Vanangamudi, K.: *Principles and Methods of Plant Breeding*, International Book, Delhi, 2005.

Whealy K.: *The Garden Seed Inventory*, Decorah, Seed Saver Publications, 1988.

White, G.F.: *Natural Hazards: Local, National, Global*, Oxford University Press, New York, 1974.

Woolfe Jennifer A.: *The Potato in the Human Diet*, Cambridge, Cambridge University Press, 1989.

Yadav, M.: *Nutritional Biochemistry and Metabolism*, Arise Pub, Delhi, 2008.

Index

❑❑❑